Working Methods in Modern Science
Edited by Prof. Dr. K. Fischbeck

Disc Electrophoresis

and Related Techniques of Polyacrylamide Gel Electrophoresis

by H. Rainer Maurer Dr. rer. nat.

Max-Planck-Institut für Virus-Forschung
Tübingen, Germany

Second revised and expanded edition
With 88 figures, 16 tables, 948 literature references

Walter de Gruyter · Berlin · New York 1971

ISBN 3 11 003495 6
Library of Congress Catalog Card Number 73-164843

Printed in Germany by Rombach + Co GmbH, Freiburg i. Br.

Preface

The excellent fractionations of proteins, polypeptides and nucleic acids obtained with *disc electrophoresis* are attracting increasing attention in

biochemistry, clinical chemistry, forensic chemistry, food chemistry, toxicology, pharmacy, pharmacology, enzymology, immunology, zoology, microbiology, botany, cytology, molecular biology, etc.

This is demonstrated by the growing number of studies which make use of this primarily *microanalytical method*.

We are indebted to ORNSTEIN and DAVIS who first described the method in 1959 and who have developed its theoretical and practical aspects in subsequent years.

What is the reason for the high resolving power of disc electrophoresis? The advantages of *gels* as electrophoretic matrices compared to liquid, granular or solid media have become generally known since the introduction of starch gel and especially, polyacrylamide gel.

Both gels afford separations in continuous homogeneous systems on the basis of the *molecular sieving effect*. As a synthetic polymerization product, polyacrylamide gel is distinguished by its

chemical stability and inertia, transparency, the possibility of controlling its broadly variable structure (to produce the most diverse pore sizes), the absence of adsorption and electro-osmosis, stability to pH and temperature variations, insolubility in most solvents, analytical purity of its components as well as the good reproducibility obtained in its preparation.

Disc electrophoresis differs from continuous polyacrylamide gel electrophoresis by the introduction of discontinuities in buffer composition, pH and pore size in the gel. These discontinuities should produce the *concentrating (stacking) effect:* the samples first concentrate as thin starting zones in the upper, e. g. acid, section of the gel, then separate in the lower, e. g. alkaline, section; the upper part with large pores through which the material migrates unhindered is called the stacking or spacer gel, while the lower section with small pores is known as the separation or running gel. The method received its name from the discontinuities and – by coincidence – the discoid shape of the zones of separated samples.

The resolving power of the method therefore is based on a *combination of the concentrating and the molecular sieving effects*. These properties as well as the special equipment for the performance of the process lead to several *advantages* compared to conventional zone electrophoretic techniques:

The small sample requirement (1–100 μg), the concentration of dilute solutions of material, simple equipment (use of glass tubes), short running times (30–60 min.), a high resolving power for molecules of the most diverse sizes (molecular weights of $< 10^4$ to $> 10^6$) and shapes, good reproducibility of the separations, simple means of detection of protein and nucleic acids in general, as well as of enzymes, antigens, polysaccharide ions and radioactively labeled polyelectrolytes in particular, and others.

The *application* of the method therefore extends to a

purity analysis of preparations and chromatographic fractions, evaluation of extraction and purification methods, analysis of protein and enzyme patterns, determination of relative protein concentrations, identification of various extracts on the basis of their protein patterns, separation and identification of microgram-quantities of radioactively labeled proteins, detection of antigens, evaluation of chemical reactions of proteins and nucleic acids after chemical or physical treatment, etc.

Disc electrophoresis is not limited to analytical purposes: *preparative* (on the milligram scale) and *ultramicroanalytical* separations (on the nano- and picogram scale) have been reported.

For example, in *applied biochemistry*, the method has proved useful as a sensitive purity test. In many cases, it surpassed ultracentrifuge and conventional chromatographic and electrophoretic techniques in its resolving power; comparable sensitivities were often obtained only with immunoelectrophoretic methods. Its field of application is notable: it extends from polypeptides to ribosomes and viruses.

Its usefulness in *practical clinical chemistry* is now being recognized, primarily as a unique means to diagnose the various hyperlipemia syndromes, and also as a rapid means to diagnose hemoglobinopathy, isozyme-indicated pathology and to screen for gammopathy in serum, cerebrospinal fluid and urine. The usefulness, in routine clinical chemistry, of its full resolving capability in the interpretation of complex abnormal protein patterns, remains to be established because the available data of statistically significant tests are not yet sufficient. In *clinical research*, however (serology, hematology, immunology, enzyme pathology, etc.) its high resolving power undoubtedly represents a valuable and in many cases, indispensable, tool; this is indicated for example, by the studies of dys- and paraproteinemias. The concentrating effect of disc electrophoresis permits the direct analysis of dilute protein solutions, for example, of cerebrospinal fluid and urine: The denaturing influences encountered during the concentration of dilute protein solutions by customary methods are thus avoided. On the whole, however, the analytical possibilities offered by discontinuous polyacrylamide gel electrophoresis have not yet been fully exhausted and explored.

The present monograph offers a summary of methodic experiences with disc electrophoresis and related techniques of polyacrylamide gel electrophoresis as far as this is possible. Moreover, it presents examples of research results obtained with this method; these should only serve as stimuli and demonstrate the diverse applicabilities of the technique. Such a "catch-all" review cannot be considered complete in view of our present-day rapid development of clinical chemistry and biochemistry. The author therefore will gratefully accept any important additions, informations, practical experiences, corrections and suggested modifications. The theoretical fundamentals have been kept brief; references are made to more detailed presentations. In the examples of application in clinical chemistry and biochemistry, it was necessary to limit the selection. Moreover, some overlapping in the description of biochemical and clinical-chemical applications was unavoidable.

The text occassionally refers to *manufacturing firms* in order to facilitate the search for sources of supplies for the reader. A list of firms appears on p. 211. It was not always possible to list comparable products of several firms. Mention of an instrument or a product therefore does not represent a quality rating.

Since the first German edition has appeared in early 1968, a great number of papers have been published dealing with the method. It is impossible to cite all of them, particularly in view of the existence of highly useful information sources such as the Canalco Disc Electrophoresis Information Center (firm 5), several handbook chapters and reviews to which the reader is explicitly referred.

Besides the early account of a conference on gel electrophoresis [W 14], which was held by the New York Academy of Sciences in late 1963 and which covered the basic principles and first applications of the methods, the excellent monograph by GORDON [G 23] and the highly informative and useful expert chapters in a volume edited by SMITH [S 53] merit particular attention.

However it was felt that each user of the book would be predominantly interested in a comprehensive survey on most methodic developments hitherto known. Such a review might enable him to select the most suitable technique to solve his particular problems. This book is primarily intended to assist him in this search, to provide him with a useful guideline for laboratory practice and to save him a time-consuming literature study and technical errors.

H. R. M.

Tübingen (Germany), June 1971

Acknowledgements

This book could not have been written without the generous assistance by the Canalco Disc Electrophoresis Information Center, Rockville (Md.), USA, which made available to me numerous data, details, theoretical and technical informations of its extensive bibliography, index and information service. I am much obliged to Mr. RALEIGH HANSL, Jr., President, and to Mr. N. CLAUDY, Information Director, for continuing and unfailing support and counsel.

My gratitude is due to the many authors who sent reprints of their papers. In particular I wish to express my sincere appreciation to all authors who willingly contributed to this monograph by supplying illustrations. Thanks are also due to these authors and the publishers of journals for granting permission to reproduce the instructive illustrations. Many experts of polyacrylamide gel electrophoresis critically discussed with me several aspects of the technique and made valuable suggestions. I am especially indebted for excellent advice to Drs. R. C. ALLEN, D. GRÄSSLIN, U. GROSSBACH, K. FELGENHAUER, S. HJERTÉN, U. E. LOENING, S. MROZEK, V. NEUHOFF, C. PELLING, H. STEGEMANN, H. TICHY and last but not least to Dr. W. THORUN, who competently contributed to the chapters on the molecular sieving effect and on molecular weight determination methods. I also thank Dr. D. M. FAMBROUGH who first introduced me to the method in Prof. J. BONNER's group at the California Institute of Technology.

For the task of typing several parts of the manuscript and collecting references I owe particular gratitude to Miss G. PAULDRACH. Finally, sincere thanks are expressed to my dear wife for her untiring assistance in the preparation of the many successive versions of the manuscript.

H. R. M.

Tübingen (Germany), June 1971

Table of Contents

1. THEORETICAL BACKGROUND

1.1. The Polyacrylamide Gel

1.1.1. Formation and Structure of the Gel

Polyacrylamide gel is the *polymerization and cross-linking product* of the monomer *acrylamide*, $CH_2=CH-CO-NH_2$, and a cross-linking comonomer, usually *N,N'-methylene-bis-acrylamide (Bis)*, $CH_2=CH-CO-NH-CH_2-NH-CO-CH=CH_2$ [B 39, O 12, R 16]. The three-dimensional network of the gel is formed by cross-linking of polyacrylamide chains growing side-by-side by the mechanism of vinyl polymerization (copolymerization in solution). This leads to the development of numerous, random polymer gel coils (Fig. 1 A) in which the polyacrylamide chains assume a state of maximum entropy, i. e. the most irregular shape. The growing coils move together (Fig. 1 B) and are cross-linked by main valencies (Fig. 1 C), where bifunctional compounds, such as N,N'-methylene-bis-acrylamide, are built into the polymer chains as cross-linking agents and can react with free functional groups at terminals of other chains. The chemical structure of the gel is shown in Fig. 2. (For other cross-linking comonomers see page 60.)

The concentration of monomer and comonomer in the gelating solution and the *degree of polymerization* (chain length) and *cross-linking* (i. e. the quantity of built-in cross-linker) determine the density, viscosity, elasticity and mechanical strength of the gel [O 9, O 12].

Gel density or *gel concentration* may be defined by two numerals T and C, where the first numeral T denotes the total percentage concentration of both monomeres (acrylamide and Bis) and the second numeral C the percentage concentration of the cross-linker relative to the total concentration T. HJERTÉN [H 23] introduced the following equations to calculate the gel composition:

$$T = \frac{a + b}{m} \cdot 100 \, [\%] \quad (1)$$

$$C = \frac{b}{a + b} \cdot 100 \, [\%] \quad (2)$$

a = acrylamide (g)
b = N,N'-methylenebisacrylamide (g)
m = volume of buffer (ml)

The *weight ratios of acrylamide to Bis* are highly critical [O 12]: If they are smaller than 10, the gels become brittle, rigid and opaque; if they exceed 100, 5 % gels (relative to acrylamide) are pasty and easily break down.

Elastic and completely transparent gels are obtained with ratios of about 30 in which the acrylamide concentration must be higher than 3 %. DAVIS [D 7] investigated the concentration ranges of acrylamide between 1.5

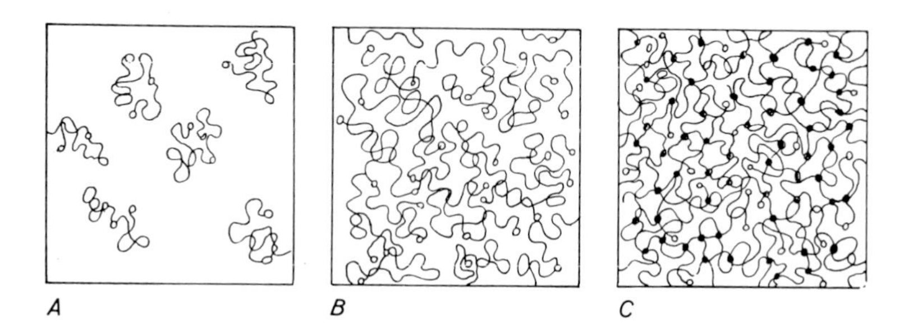

Fig. 1 – Schematic representation of the formation of polyacrylamide gel from random gel coils. Transition from a dilute polymer solution (A) through the concentrated solution (B) to the gel (C). –o– cross-linking agent; –•– tie-points. From VOLLMERT [V7].

```
          |
          CH2
          |
          NH
          |
          CO
          |
 -CH2-CH-[CH2-CH-]xCH2-CH-[CH2-CH-]xCH2-
                  |        |        |
                  CO       CO       CO
                  |        |        |
                  NH2      NH       NH2
                           |
                           CH2
                           |
                           NH
                           |
                           CO
                           |
 -CH2-CH-[CH2-CH ]xCH2-CH-[CH2-CH-]xCH2-
       |        |                |
       CO       CO               CO
       |        |                |
       NH       NH2              NH2
       |
       CH2
       |
       NH
       |
       CO
       |
 -CH2-CH-[CH2-CH-]xCH2-CH-[CH2-CH-]xCH2-
                  |        |        |
                  CO       CO       CO
                  |        |        |
                  NH2      NH       NH2
                           |
```

Fig. 2. – Structure of polyacrylamide gel. Monomer: acrylamide; comonomer (cross-linking agent): N,N'-methylene-bis-acrylamide. From OTT [O 12].

and 60 % and of Bis between 0 and 0.625 %. He found that no further gelation takes place at concentrations of less than 2 % acrylamide and 0.5 % Bis; furthermore, he noted that an increase of the acrylamide concentration should be generally accompanied by a decrease of the Bis concentration in order to produce elastic gels. RICHARDS et al. [R 34] proposed the following "compromise" formula for this requirement:

$$C = 6.5 - 0.3\,T \tag{3}$$

This formula can be used to calculate the composition of manageable gels within the range of 5–20 %; the value of C is not critical and may be varied by about ± 1 % in most cases. Taking optimal resolution of human serum proteins as a basis, BRACKENRIDGE and BACHELARD [B 53] calculated the optimal cross-linkage for a given gel percentage using the least squares regression equation:

$$B = 0.201 - 0.0112\,T \tag{4}$$

where B is the Bis concentration in % and T as above.

1.1.2. Polymerization: Catalysts and Processes

Usually certain *catalyst-redox systems* which furnish free radicals are used for the polymerization of polyacrylamide gels for electrophoresis [L 38], for example:

Ammonium persulfate	– N,N,N',N'-tetramethylethylenediamine (TEMED)
Ammonium persulfate	- 3-dimethylaminopropionitrile (DMPN)
Ammonium persulfate	– DMPN-sodium sulfite [S 66 a]
Hydrogen peroxide	– ferrous sulfate - ascorbic acid [J 17]
Riboflavin	– TEMED (as a photocatalytic system).

Even small concentrations of tertiary aliphatic amines, such as TEMED, accelerate the polymerization, while DMPN and triethanolamine appear to be less active [D 7]. Ammonium persulfate has the following advantages: it can be prepared in high purity, is relatively stable at 0° C and has little tendency to develop molecular oxygen. Suitable catalytic systems must not change the selected buffer conditions, electrical conductivity and viscosity of the gels.

The *chemical polymerization* is initiated from free monomer radicals which in turn are produced by the base catalyzed formation of free oxygen radicals from *persulfate*. Since the free bases of TEMED or DMPN are needed, polymerization may be delayed or prevented at lower pH (see also page 55). However, reproducible gels are only obtained if the polymerization is not retarded. Molecular oxygen, cooling and impurities (e. g. metals) can inhibit or even prevent chemical polymerization [0 12]. Consequently, it is usually advisable to select and standardize the conditions such that the solutions will gelate within about one hour.

The initial *rate of polymerization* was found by UV absorbance measurement to be proportional to the square root of the concentration of *ammonium persulfate* [W 4a]. At pH 8.8 a 7.5% solution of acrylamide shows a lag period of several minutes, then an abrupt increase of the reaction rate which declines as rapidly and levels off after about 30 min. At pH 4.3 the initial rate of polymerization is much slower, hence about 90 min. are required for completion of polymerization. Addition of 5 mM sodium sulfite reduces the lag period to about 2 min.

Polymerization of highly concentrated or large diameter gels produces considerable heat which may lead to cavities in the gel from dissolved gases and to separation of the gel from the wall. The effect can be diminished by decreasing the temperature of the gel solution to about 6° C prior to polymerization [P 15].

In contrast to the chemical polymerization by persulfate the *photochemical polymerization* with *riboflavin* requires traces of oxygen to take place [G 23]. Photodecomposition of riboflavin, with the consequence of reoxidation of leucoflavin and production of free radicals, can only occur if molecular oxygen is initially present. Excess of oxygen, however, may limit chain growth and should be avoided.

Possible artifacts arising from the interaction of sample proteins with the catalysts persulfate and riboflavin are discussed on pages 59 and 60.

Needles [N19a] has discussed the effect of several solution additives (e. g. Tris, TEMED, HCl, Glycine) on *photopolymerization* and has suggested that polymerization accelerators (like TEMED) are generally mild reducing agents (hydrogen donators) which may complex with riboflavin and/or monomers through donor-acceptor interactions.

The starting monomers can also be dissolved in glycerol, ethylene glycol, concentrated urea or saccharose solutions instead of water without a notable inhibition of gelating power; however, a reduction of the gelation time must be anticipated with an increasing viscosity of the solvent [R 8]. Urea can be incorporated in polyacrylamide gel in a concentration of 8 M at acid [E 1] and alkaline [C 47] pH values.

1.1.3. The Molecular Sieving Effect in Polyacrylamide Gel

By W. Thorun and H. R. Maurer

1.1.3.1. Introduction

In conventional electrophoresis using liquid media, electrically charged particles are separated mainly according to their net charge. In paper electrophoresis, for example, the paper merely serves as a supporting and anticonvection medium which exerts little influence on the separation process itself. *Gels*, however, are distinguished from liquid media by high viscosities and high frictional resistances. These supporting media not only prevent convection and minimize diffusion, but also actively participate in the separation process by interacting with the migrating particles. This interaction depends on the particle size. Consequently, in gels the particles are separated according to both charge *and* size. The property of gels to distinguish molecular species of different sizes is ascribed to their *size sieving capacity*. Therefore this phenomenon has been termed "*molecular sieving effect*".

For example, in starch gels the mobilities of hemoglobin-haptoglobin complexes are always lower than those of the individual proteins, while in paper electrophoresis, where molecular sieving effects do not occur, the complexes show mobilities intermediate between those of their components.

Molecular size sieving does not play the same role in all types of gels: while it predominates in *starch* and *polyacrylamide gels*, it is only small in *agar gel* [R 11]. Moreover, in the presence of alkaline buffer solutions, agar gel, as an acidic polysaccharide, exhibits electro-osmotic currents with their unfavorable influence on several separation problems. Starch gel shows no such effects. Yet the succes of starch gel electrophoresis is highly dependent on the quality of the starch gel, which may not always be prepared from natural products with uniformly good reproducibility. This shortcoming does not exist for polyacrylamide gel as a synthetic polymer: under constant conditions and with chemically defined components of relatively high purity, the gel can always be prepared in a reproducible manner. Moreover, the density of the gel network (pore size) can be varied in a wider range than is possible for starch gels: The gel concentration (referred to the monomers) can range between about 2 and >30 % and may thus be readily adjusted to most separation problems.

1.1.3.2. Relations between Gel Density (Pore Size), Cross-Linkage, Particle Size and Electrophoretic Migration

Attempting to explain the retardation of a particle by a gel network, SMITHIES [S 58] and ORNSTEIN [O 9] used the terms "*pore*" and "*pore size*". As the migrating particle must tunnel through the pores, the particle is strongly retarded, when the pores are small in a highly concentrated gel; in contrast, the particle is only slightly hindered, when the pores are large in a lowly concentrated gel.

On the basis of this *pore concept*, one can calculate average pore sizes and the statistical distribution of pore sizes of a gel. However, different theoretical approaches are possible, thus leading to different relations between mean pore size and gel density. One approach was reported by ORNSTEIN [O 9] who calculated, that a 7.5 % polyacrylamide gel should have a mean pore diameter of 50 Å and a 30 % gel a mean pore diameter of about 20 Å. In another approach RAYMOND and NAKAMICHI [R 13] proposed a formula by which the average pore diameter $\bar{p}$ is inversely proportional to the root of a volume concentration c of the polymer having a molecular diameter d:

$$\bar{p} = \frac{kd}{\sqrt{c}} \qquad (1)$$

where the volume concentration c of the gel is defined by:

$$c = k^2 d^2 / \bar{p}^2 \quad [\text{d and p in Å}] \qquad (2)$$

The factor k depends on the geometric configuration of the gel; it amounts to 1.5 if one assumes that the polymer strands are cross-linked at approximately right angles. In a 5 % polyacrylamide gel with uncoiled chain strands (5 Å diameter) a pore diameter of 38 Å may thus be calculated.

Another calculation by TOMBS [T 18] relates pore size, gel concentration and diameter of cylindrical strands of which the gel is presumably composed:

$$\bar{p} + d = \frac{\sqrt{3\pi}\, 5\, d}{\sqrt{c}} \qquad (3)$$

Numerous and elaborate calculations of pore sizes were also performed in gel chromatography [e. g. A 4, L 7 a, O 2 a].

Opinions differ whether these calculations would allow to predict the electrophoretic mobility of a particle on the basis of the computed gel pore size and the particle dimensions. TOMBS [T 16, T 18], INGRAM et al. [I 1] and recently RODBARD and CHRAMBACH [R 48 a] have found relations between pore size, gel density and particle mobility. On the other hand, the question may be raised, whether the pore size concept is the only way towards a correct explanation of the gel electrophoretic phenomena. Apparently some authors [H 15, H 25, T 11] do not favour the pore concept. For example, HJERTÉN [H 25] prefers to speak of "*size sensitivity*" and HEDRICK and SMITH [H 15] cautiously expressed that "the words sieving action, sieved are . . . not to be equated with the physical processes responsible for these phenomena."

The pore concept should therefore be regarded only as a working hypothesis, which has proven fruitful in many respects, but not in all. For example, it can neither explain the strong dependence of migration on the cross-linkage of the gel (see page 9), nor the fact, that molecular sieving effects are also observed in the ultracentrifuge with viscous solutions as hyaluronic acid [L 7]. Moreover, almost every pore theory tacitly assumes that a gel is composed of a matrix of rigid, cross-linked gel strands, which are not displaced by the migrating particle. As pointed out by GORDON [G 23], however, proteins show rather slow diffusion rates in gels through which they have been passed rapidly by gel electrophoretic forces. This suggests a gel network of elastic gel strands, which can be bent by the migrating proteins. Hence pore size measurements on the basis of diffusion data may be only of limited value.

Recent studies stress the remarkable influence of the cross-linkage on the electrophoretic migration [B 53, F 10, H 15, J 16, R 48a, T 11, Z 10]. To account for this influence, a cross-linkage coefficient k_C can be defined (eqn. [2], page 9) which is linearly related to the root of the cross-linkage C (Fig. 3) for each specified gel preparation made from equal reagents under identical conditions of polymerization [T 11]. Moreover, at constant cross-linkage the coefficient may still depend on the particular gel preparation, indicating variations in the molecular sieving capacity due to chemicals only partially reactive or conditions preventing complete polymerization. For example, different coefficients were obtained when gels were made from different acrylamide batches. The coefficient may hence be used as a criterion to control and improve the reproducibility of gel formation.

These findings do not support previous statements of ORNSTEIN [O 9] and RAYMOND and NAKAMICHI [R 13], who believed that the cross-linking agent Bis only provides for mechanical stability of the gel. These authors based their concept on diffusion and permeability tests performed with relatively small molecules by WHITE et al. [W 16–17], which were then subjected to criticism [J 16, H 15].

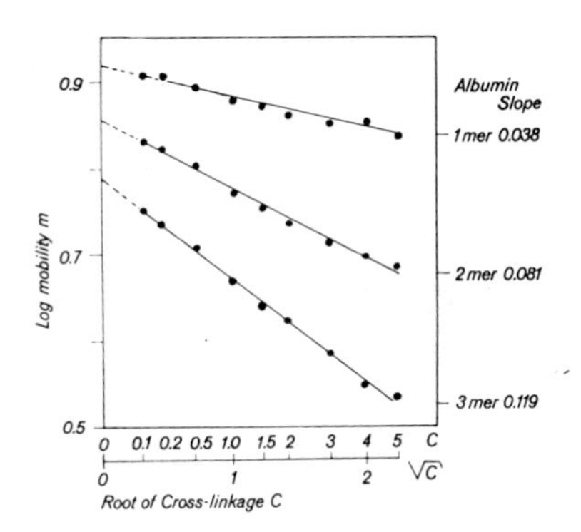

Fig. 3. Linear relationship between log mobility and the root of cross-linkage C at constant gel density (T = 6.5 %). Extrapolation of the lines to C = 0% gives approximate values of the mobilities in a 6.5% polyacrylamide solution. From: THORUN [T 11].

In *gel chromatography*, where the pore theory seems to be widely accepted, pore sizes were found to be dependent on the cross-linkage, too. Using granulated gels in chromatographic sieving studies, FAWCETT and MORRIS [F 10] determined a minimum pore diameter at C = 5 % for any given T. A gel of greater T may therefore show larger pores than one of smaller T, if C is above or below C = 5 %. Similar findings with electrophoresed gels were reported by HJERTÉN et al. [H 28]. FAWCETT and MORRIS [F 10] constructed curves relating mean pore size to C for gels of T = 6.5–20 %. See also [W 15].

Whatever conclusion one may draw, either that the degree of cross-linkage effects migration behaviour in some yet undetermined way, or by first effecting the pore size secondly influences migration, the cross-linkage is an important parameter for gels. To obtain reproducible and comparable results in different experiments, it is therefore essential to keep it at a constant value throughout all experiments. Moreover, variable factors of gel formation, such as catalyst concentration and gelling time, may greatly affect relative mobilities [K 19a].

1.1.4. Molecular Weight Determination by Polyacrylamide Gel Electrophoresis

1.1.4.1. Proteins

By W. Thorun and H. R. Maurer

Evidence for the size-sensitivity of gels was obtained at approximately the same time in *gel electrophoresis* (Smithies 1955 [S 57]) and in *gel filtration* (Lathe and Ruthren 1956 [L 6a]). Subsequently gel chromatography progressed much quicker than gel electrophoresis. Calibration curves were constructed relating the elution volume and other parameters with molecular size. Today molecular weight (MW) determination by gel chromatography is a routine.

In contrast, MW determination by gel electrophoresis is still in a state of test, but the reported results are encouraging. A definite advantage of gel electrophoresis is its preciseness due to its high resolution power. Moreover, it also gives very accurate results, when strongly denaturating solvents are used, which transform the molecules into a random coiled state, thus allowing for example to determine the MWs of membrane proteins which are in general not readily soluble in aqueous buffers. Experiments investigating migration behaviour in relation to molecular size or gel density have been reported by many authors [D 12a, D 39, F 14a, H 15, I 1, K 31a, M 66, P 5a, R 13, R 15, R 48a, S 34–35, S 58, T 10–11, T 16, T 18, W 5, W 25a, Z 7].

In gel electrophoresis separation takes place due to both charge and size differences of the molecules. The influence of charge and the difficulties to eliminate it may have been a handicap towards a method for MW determination by gel electrophoresis. As to the *elimination of charge*, the above cited papers can be divided into *two categories*. In the first category, electrophoresis is performed in gels of different densities and the quotient of the migration distances (or retardation quotient) is measured which is completely independent of charge because in such a quotient the charge is cancelled. In the second category, electrophoretic conditions (solvent, gel density) are chosen such that charge differences are presumably so minimal that they may be neglected.

In the following some of the essential points of these papers will be presented. Confusion may arise from the fact that these studies not only differ in their results to some extent but also in the terminologies preferred by the authors. Therefore to facilitate comparison a uniform terminology was introduced.

Table 1 lists the proposed formulae of the *first category* in the left column in their original form, in the right column in a more uniform terminology. In this way it can better be evaluated where there is consensus in the findings of different authors and where the results are in contrast.

There is general agreement among most authors [F 14a, H 15, M 66, R 48a, P 5a, T 11] that the logarithm of the electrophoretic mobility m of a particle is linearly related to the gel density T:

$$\log m = \log m_0 - K \cdot T \quad (1)$$

where m_0 is the free mobility (at 0% "gel" density) and K the retardation constant. The mobility can either be measured as migration distance *d* obtained under a constant voltage gradient *E* and constant electrophoresis time *t*: $m = d/t \cdot H$ (see page 20), or as migration distance divided by that of the buffer front (R_m-value) [H 15].

In Fig. 4 a plot of *log m* versus *T* is shown as obtained by THORUN [T 11] with polymers of cristalline serum albumin. Other authors [F 14a, M 66, H 15, R 48a, P 48a] have constructed similar plots. As the ordinate intercepts of the straight lines should correspond to $log\ m_0$ of each polymer, the free mobilities m_0 can be extrapolated realizing a suggestion by ORNSTEIN [O 9].

The nature of the retardation constant *K* in eqn. (1) was investigated by several authors [H 15, R 48a, T 11]. On the basis of a complex theoretical model, RODBARD and CHRAMBACH [R 48a] suggested two equations relating the particle radius, the gel fiber radius and *K* (Table 1). By electrophoresing a series of test proteins in gels of increasing densities but constant cross-linkage, and by forming plots like in Fig. 4 for every protein, other authors [H 15, T 11] found a linear relationship between *K* and *MW*. For example, when the slopes of the lines in Fig. 4 are plotted versus the MWs of the albumin polymers, like in Fig. 5, a straight relationship is gained [T 11]:

$$K = k_C \cdot MW + a \tag{2}$$

where *a* is the ordinate intercept of the calibration line and k_C the cross-linkage coefficient which has a constant value for a constant cross-linkage (see page 6).

It follows from equations (1) and (2):

$$\log m = \log m_0 - (k_C \cdot MW + a) \cdot T \tag{3}$$

This equation describes the electrophoretic mobility *m* of a compact spherical particle in a gel of density *T* and cross-linkage *C* as a function of its size (proportional to MW) and free mobility m_0, which in turn depends on its effective charge *Q* ($m_0 = Q/6\pi r\eta$, see page 20). Equation (3) can be modified to:

$$\log (m_1/m_2) = (k_C \cdot MW + a)(T_2 - T_1) \tag{4}$$

where m_1/m_2 is the quotient of the mobilities of a single protein in two gels of the densities T_1 and T_2 ($T_2 > T_1$).

Table 1: The molecular sieving effect in polyacrylamide gel electrophoresis:
Dependence of electrophoretic migration on parameters of the gel and of the particle to be separated.
From: THORUN [T 11].

Author(s)	Ref.	Equation by author(s)	Equation in uniform terminology	Nomenclature
TOMBS	[T 16]	$MW = K\left(\frac{1}{C}\right)^3_{M/D=0.5}$	$MW = \frac{const.}{T^3_{0.5}}$	T = gel density (% total monomer concentration) $T_{0.5}$ = gel density required to reduce m_0 to half its value C = cross-linkage k_C = cross-linkage coefficient MW = molecular weight K = retardation constant m = electrophoretic mobility in gel of density T m_0 = free mobility (mobil ty in "gel" of density 0) a = constant R = particle radius r = gel fiber radius
INGRAM et al.	[I 1]	$P_L = k_1 + k_2 \left(\frac{1}{C}\right)$	$2r = const. + \frac{const.}{T_{0.5}}$	
MORRIS	[M 66]	$\log M = \log M_0 - KT$	$\log m = \log m_0 - K \cdot T$	
ZWAAN	[Z 7]	$\log M = a(r) + b$	$\log MW = const. \frac{m_2}{m_1} + const.$ $(T_2 > T_1)$	
THORUN	[T 11]	$\log m = \log m_0 - (k_C \cdot MW + a) \cdot T$ $\log (m_1/m_2) = (k_C \cdot MW + a)(T_2 - T_1)$ $(T_2 > T_1)$		
RODBARD and CHRAMBACH	[R 48a]	$\sqrt{K_R} = c_1 (\bar{R} + r)$ $\sqrt[3]{K_R} = c_2 (\bar{R} + r)$	$\sqrt{K} = const. (R + r)$ at $C < 5\%$ $\sqrt[3]{K} = const. (R + r)$ at $C > 5\%$	

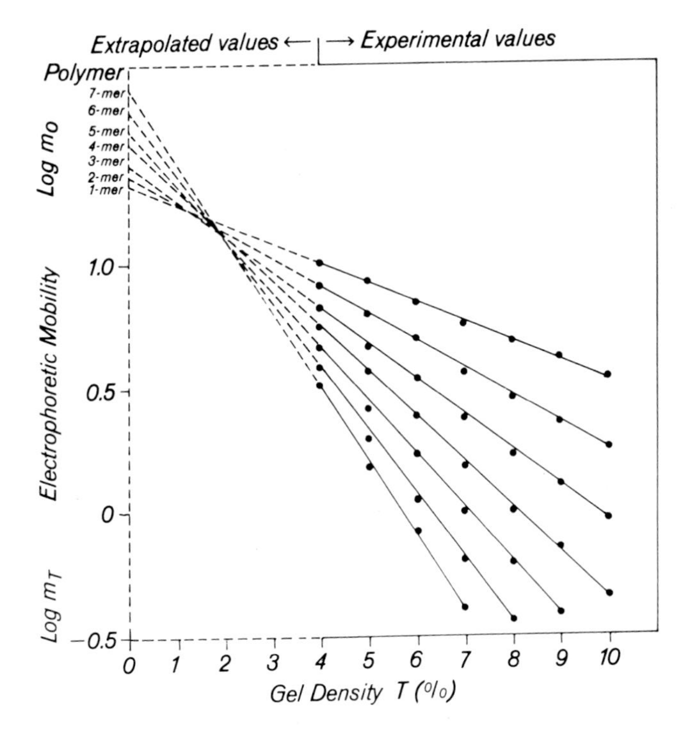

Fig. 4. Plot of the logarithm of the electrophoretic mobility versus gel density. From: THORUN [T 11].

Crystalline serum albumin was electrophoresed across a gel gradient (C = 2% throughout) and the distances of migration from the first up to the seventh polymer were measured at the corresponding gel densities T. The straight lines were calculated from the data by means of a least squares computer program. The ordinate intercepts of the lines are equal to log m_0 (eqn. 2). The slopes of these lines are linearly related to the MWs (eqn. [3] and Fig. 5).

In practice, for the determination of MWs, eqn. (4) can be written in a simplified form:

$$\log m_1/m_2 = k_1 \cdot MW + k_2 \tag{5}$$

To calculate, for example, the MW of hemoglobin, we take the following experimental data from ZWAAN's work [Z 7]:

Protein	Migration distance d [mm] 4% gel d_1	log d_1	12% gel d_2	log d_2	log d_1/d_2 = log m_1/m_2	MW × 10^{-3}
Transferrin	110	2.041	17	1.230	0.811	90
Ovalbumin	240	2.380	68	1.833	0.547	45
Hemoglobin	131	2.117	29	1.462	0.655	?

By fitting the values of the last two columns into eqn. (5) the following three equations with three unknowns are obtained: $0.811 = k_1 \cdot 90 \cdot 10^3 + k_2$; $0.547 = k_1 \cdot 45 \cdot 10^3 + k_2$; and $0.655 = k_1 \cdot MW_{Hb} + k_2$. From these a MW of 63.4×10^3 can be calculated. It follows from this example that such a calculation is only permissive if two reference proteins of similar partial specific volume as the protein under test are electrophoresed under identical conditions in the same run.

ORNSTEIN [O 9] already suggested to measure mobility quotients, which are independent of charge, in order to determine the free mobilities and the aqueous diffusion constants of different proteins. RAYMOND and NAKAMICHI [R 13, R 15] then used the method in their Orthacryl technique for comparison of relative MWs and

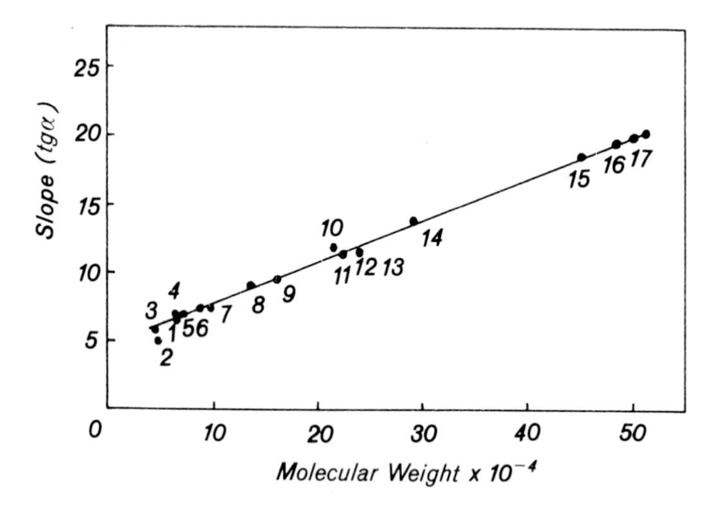

Fig. 5. Plot of the slope of log mobility lines (Fig. 4) versus the molecular weight of test proteins. From: HEDRICK and SMITH [H 15].

The proteins used are designated by numbers: 1 pepsin dimer, 2 ovalbumin, 3 α-amylase, 4 albumin monomer, 5 transferrin (human), 6 ovotransferrin, 7 hexokinase, 8 lactate dehydrogenase, 9 ketose-l-phosphate aldolase, 10 β-amylase, 11 nicotinamide deamidase, 12 α-urease, 13 catalase, 14 xanthine oxidase, 15 apoferritin (ferritin) monomer, 16 urease, 17 ribulose diphosphate carboxylase.

ZWAAN [Z 7] to determine MWs of water soluble proteins. The method was extended to phenolic solvents by THORUN and MEHL [T 10] for MW determination of nerve membrane proteins and applied by DEMUS and MEHL [D 12a] for the simultaneous MW determination of 14 erythrocyte membrane proteins.

ZWAAN's [Z 7] empirical equation (Table 1) relates a *retardation quotient* linearly to a log MW (Fig. 6). Zwaan defines this retardation quotient as the ratio of the mobility in a gel of higher density to the mobility in a gel of lower density; it is therefore inversely proportional to the above mobility quotient. To correlate ZWAAN's data with eqn. (4), THORUN [T 11] plotted the log of the inverse retardation quotients of Fig. 6 versus the corresponding MWs. The obtained linear relationship (Fig. 7) support either eqn. (4) or ZWAAN's equation (Table 1).

TOMBS' [T 16] equation (Table 1) connects the radius of the particle with the gel density $T_{0.5}$ required to reduce the free solution mobility m_0 to half its value. TOMBS derived his eqn. from an estimate of mean pore sizes of gels, making certain assumptions on how this pore size would possibly effect migration. INGRAM et al. [I 1] suggested a similar equation. However these relations do not agree with recent findings [R 48a, T 11, Z 7]. For more recent studies on gel and particle parameters see [B 35a, F 13b, P 5a, R 48a].

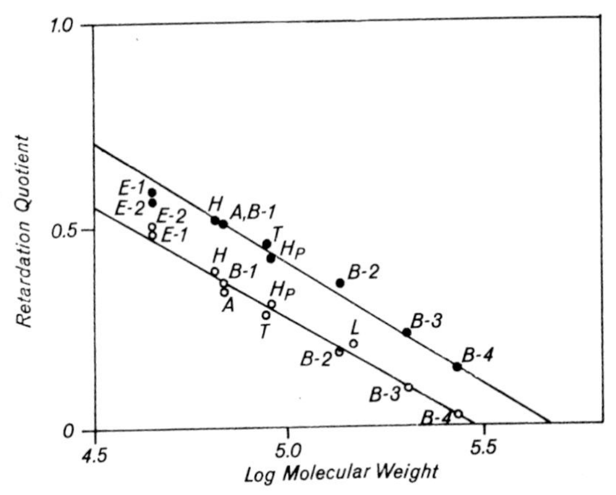

Fig. 6. Linear relation between log molecular weight of standards and ratios of corrected absolute mobilities (or retardation quotient) in 2 different gel concentrations (8 and 5 %, closed circles, and 10 and 15 %, open circles). Combination of 3 runs in each gel concentration. From: ZWAAN [Z 7].

Proteins used: E–1, E–2: Chicken ovalbumin 1 + 2, H: Human hemoglobin A–1, A: Human serum albumin, T: Human transferrin, Hp: Human haptoglobin 1–1, L: Human lactate dehydrogenase H, B 1–4: Borine serum albumin mono– and polymers.

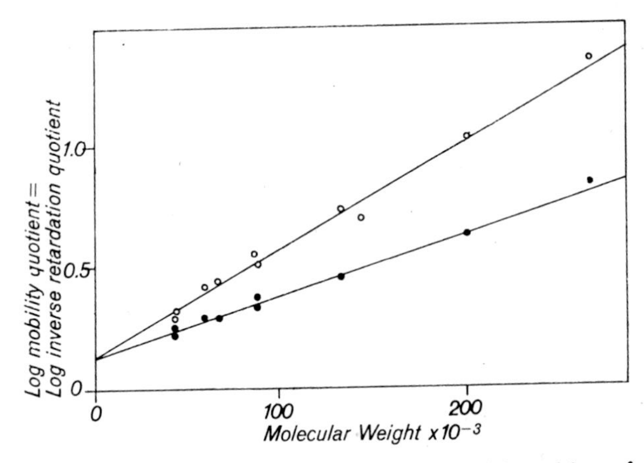

Fig. 7. Linear relation between molecular weight and logarithm of the inverse retardation quotient of ZWAAN [Z 1] (eqn. 4). From: THORUN [T 11].

All experimental data were obtained from Fig. 6, but the plot is completely different. Thus every open or closed circle in Fig. 6 has a corresponding circle in this figure.

In analogy to gel chromatography the reliability of MW determination by gel electrophoresis clearly depends on the extent to which different proteins conform to a common relationship between MW and migration behaviour. Therefore less precise results are to be expected for molecular species with partial specific volumes significantly different from 0.73 (ml/g) or non-spherical shapes. In addition, carbohydrate and lipid content as well as prosthetic groups may probably cause deviations. More information is needed to evaluate the influence of these parameters.

For example, there are indications that the elongated shape of bovine fibrinogen is responsible for its anomalous migration behaviour [T 10, T 11].

Disturbances by differences in partial specific volume and shape should be reduced when the proteins are electrophoresed in a medium favouring the *random coiled state*. According to THORUN and MEHL [T 10] a mixture of *phenol-acetic acid-water* (2 : 1 : 1, w/v/v) has this property. In addition,

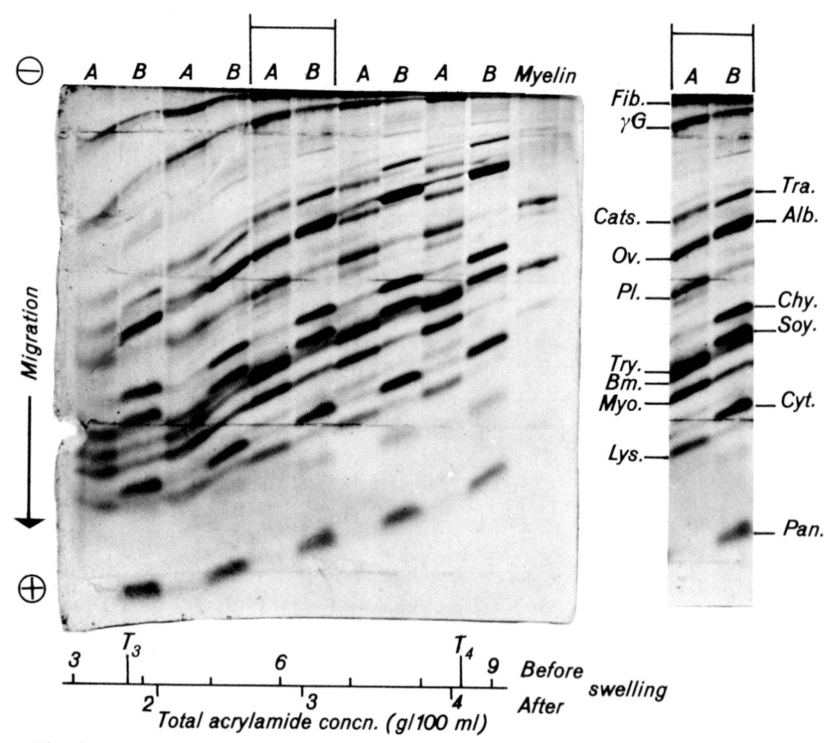

Fig. 8. Electrophoresis across a gel gradient in phenol-acetic acid-water (2:1:1). From: THORUN and MEHL [T 10].

Molecular weights × 10^{-3} of the reference proteins: Sample A: fibrinogen (Fib.), 340; human γ-globulin (γG), 160; catalase (in this solvent as subunits) (Cat.), 59; ovalbumin (Ov.), 45; Trypsin (Try.), 23.8; Myoglobin (Myo.), 17.8; Lysozyme (Lys.), 14.5. Myelin proteins: proteolipid protein (Pl.), basic myelin protein (Bm.). Sample B: transferrin (Tra.), 90; albumin (Alb.), 67; Chymotrypsinogen (Chy.), 25; soybean trypsin inhibitor (Soy.), 22.6; cytochrome *c* (from horse) (Cyt.), 16.5; pancreatic trypsin inhibitor (Pan.), 9.

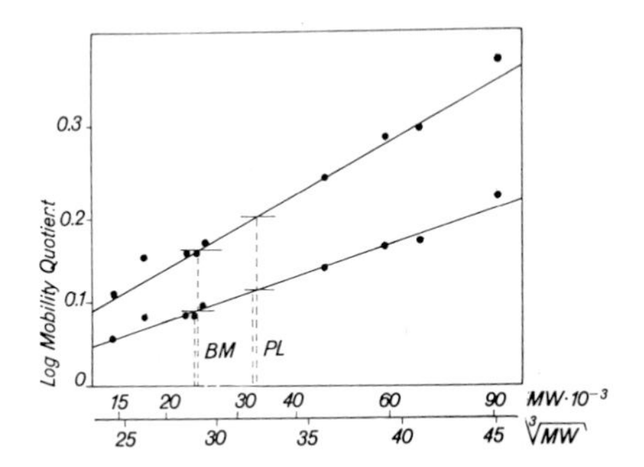

Fig. 9. Calibration plot of mobility quotients versus the third root of the molecular weights of random coiled proteins (eqn. [6]). From: THORUN [T]

The data were obtained from Fig. 8 at gel densities T_1 and T_2 (circles), and T_3 and T_4 (dots). The calibration line was computed by the method of least squares from 10 points excluding myoglobin, with a mean single deviation of $\pm$ 5 %. BM: Basic myelin protein, PL: proteolipid protein.

this system is able to solubilize membrane proteins. These authors determined the MWs of a basic myelin protein and a proteolipid of nerve membranes (Fig. 8). With this solvent, however, a linear relationship between the logarithm of the mobility quotient and MW, as in eqn. (4), is not obtained but there is a satisfactory linear relation between the logarithm of the mobility quotient and the third root of the MW (Fig. 9):

$$\log (m_1/m_2) = \text{const.}_1 + \text{const.}_2 \cdot \sqrt[3]{MW} \quad (6).$$

By testing the validity of either eqn. (4) or (6) one should therefore be able to decide whether the particles are compact or random coiled in a specific solvent.

Similar studies [P 5a] with 8 M urea as dissociating solvent have been reported.

While these studies of the first category completely eliminate the influence of charge, the following approaches of the *second category* choose solvent and gel conditions such that charge differences are presumably so minimal that they may be neglected. This may not be always permissive, hence the influence of charge should be kept in mind. Several authors consider these methods to be simpler and faster than those of the first category, since only one unidimensional run is required.

SHAPIRO et al. [S 34–35] incorporated *sodium dodecylsulfate* (*SDS*) in their solvent system and found an inverse linear relationship between relative migration distance and log MW, permitting a rapid and simple estimation of proteins and polypeptides (Fig. 10). These workers denatured and reduced their proteins at pH 7.1 in 1 % SDS and 1 % mercaptoethanol (ME) at 37° C for 3 hrs. Despite the choice of a group of proteins with a range of isoelectric points from 4 to 11 most points of the resulting curve fitted the relationship between MW and migration distance x:

$$MW = k\,(10^{-bx}) \quad (7)$$

where b is the slope and k a constant.

This suggests that SDS minimizes the negative charge differences and that all proteins migrate as *anions* as the result of complex formation with SDS. The extensive disruption of hydrogen and hydrophobic bonds by SDS as well as disulfide linkages by ME results in the quantitative solubilization of many relatively unsoluble proteins. In some cases, however, it may be necessary to carboxymethylate free sulfhydryls exposed by SDS to prevent aggregation during electrophoresis.

According to WEBER and OSBORN [W 5] this method permits the determination of MWs with an accuracy of less than ± 10 %. A number of proteins with MWs ranging from 1.1–7.0 × 10⁴ fitted a linear relationship after electrophoresis in gels containing 10 % acrylamide, 0.27 % Bis and 0.1 % SDS. However, when proteins of MWs from 4 × 10⁴ up to 2.2 × 10⁵ (e.g. myosin) were analyzed in gels containing half the amount of cross-linker, a hyperbolic curve was observed.

Studying the apparent influence of gel concentration on these curves in more detail, DUNKER and RUECKERT [D 39] found that each plot yields a straight line over a MW range which is characteristic of the gel concentration

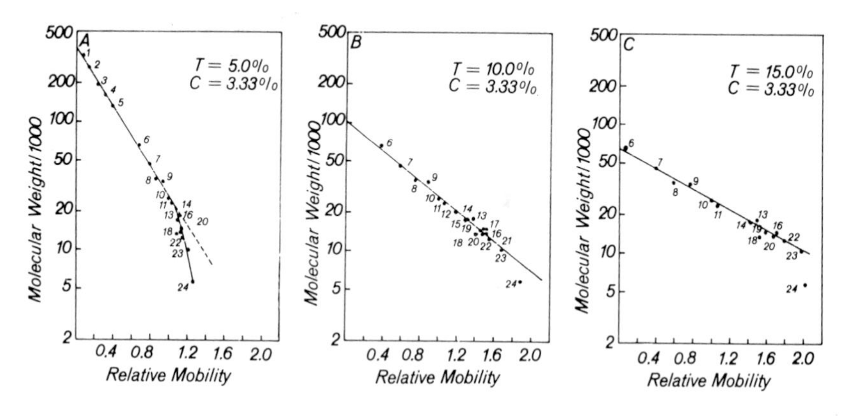

Fig. 10. Plot of the logarithm of the molecular weights of a series of proteins versus their relative mobilities on (A) 5 %, (B) 10 % and (C) 15 % gels containing 0.1 % sodium dodecyl sulfate. From: DUNKER and RUECKERT [D 39].

Relative mobility: migration distance of protein divided by migration distance of chymotrypsinogen, simultaneously electrophoresed by means of the split gel technique (see p. 64). The ordinate intercepts are specific for each gel composition. Measurements at gel ends may be inaccurate due to atypical porous zones (see p. 37). The numbers designate test proteins with MWs ranging from $3.3 \cdot 10^5$ (bovine serum albumin pentamer, No. 1) to $5.7 \cdot 10^3$ (insulin, No. 24).

(Fig. 10). Thus, a 15 % gel (C = 3.33 %) produces a linear plot for the range of 1–6 × 10^4 daltons, 10 % gels (C = 3.33 %) for 1–10 × 10^4 daltons and 5 % gels (C = 3.33 %) for 2 × 10^4 – 3.5 × 10^5 daltons. Each plot shows a critical point beyond which the MWs deviate from linearity. For 5 % gels this point corresponds to about 2 × 10^4 daltons. Moreover, it was observed that the effects of intrinsic molecular charge and conformation on electrophoretic behaviour are relatively small in the presence of SDS, in contrast to chain unfolding agents like urea.

This suggests a model in which the SDS anions form a micellar complex of definitive size with the proteins. Although size and conformation (state of foldedness) effect the stoichiometry of the complex, the interplay of anion binding and frictional resistance to gel passage produces a relatively constant log size to mobility ratio.

Using *pore limit electrophoresis* (see page 61), KOPPERSCHLÄGER et al. [K 31a] suggested a method for MW determination of proteins in linear gradient gels

1.1.4.2. Nucleic Acids

Nucleic acids show, over a wide range of molecular weights, almost constant electrophoretic mobilities in free electrophoresis at neutral or alkaline pH values [O 4], hence cannot be fractionated by this means. This has been ascribed to their *constant charge/mass ratio* [R 34, R 36]. However, media with molecular sieving properties permit the fractionation according to size and shape. Rather acid buffer systems (pH 2.3 – 3.4) even allow the separation of oligonucleotides of equal size (compositional isomers) according to charge [B 30] (see page 180).

The electrophoretic mobility of low molecular weight RNA in polyacrylamide gels is inversely related to the sedimentation coefficient [L 36, M 40, R 34]. The same is true for ribosomal RNA [L 37]. The relationship is linear over a certain range [Fig. 11] and depends strongly on the total gel concentration. RICHARDS et al. [R 34] and McPHIE et al. [M 40] found the following approximate relations:

$$s^{\circ}_{20,w} = 7.6\text{–}6.7 \cdot R_m \quad \text{for gels of } T = 10 \text{ and } C = 5\%,$$
$$s^{\circ}_{20,w} = 13.6\text{–}11.3 \cdot R_m \quad \text{for gels of } T = 5 \text{ and } C = 5\%,$$

where R_m denotes the relative mobility (ratio of the migration distance of the RNA to that of the anion front, see page 9). For gels containing 2.6% acrylamide ($C = 5\%$) LEWICKI and SINSKEY [L 21a] found that the migration distance is a linear function of the logarithm of the Svedberg unit. Since the sedimentation coefficient, for a series of RNA species, varies as the square root of the weight – average molecular weight [B 40], it follows that the relative mobility should be inversely related to the logarithm of the molecular weight (Fig. 12 and 13).

This has been verified by several groups [B 31, L 36, P 9, R 42]. The linear relation has been found to be independent, within limits, of salt and gel concentration [L 36] and holds true for both pure polyacrylamide (Fig. 12) and composite agarose-polyacrylamide gels (Fig. 13). To determine MWs, at least two reference RNAs are required to construct a calibration curve for a given gel concentration. It appears from Fig. 13 that maximum resolution between two RNA species with similar MWs occurs when the mean MW is approximately one half of the intercept MW (M_0). Thus, to discriminate between liver 5 S and E. coli 5 S (MW around 3.5×10^4) one should choose a 10 % gel in which the M_0 is approximately 7×10^4 [P 9].

LOENING [L 36] examined the effect of *changes in conformation* and found that the decrease of relative mobility of low G-C-RNA in a low salt buffer is related to a small relative decrease in the sedimentation coefficient. However, Mg^{2+} causes a

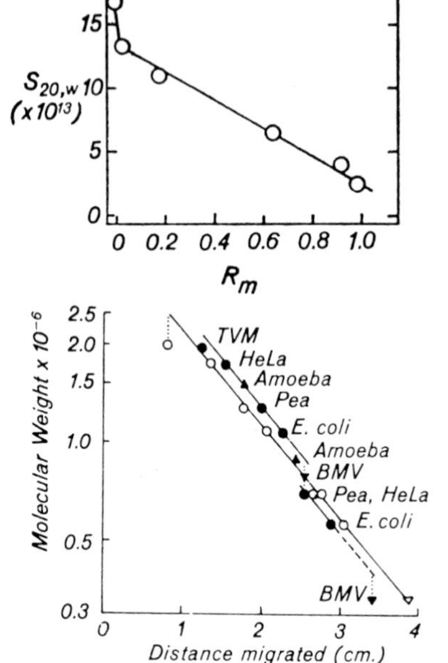

Fig. 11. Relation between sedimentation coefficient ($s_{20,w}$) and relative mobility (R_m) of different RNA species in polyacrylamide gels of T = 5 and C = 5%. From: McPhie et al. [M 40].

Fig. 12. Relation between molecular weight and electrophoretic mobility in low-salt and Mg buffers. From: Loening [L 36].

Electrophoresis of the mixture of TMV RNA, HeLa-cell r-RNA, pea r-RNA and *E. coli* r-RNA using low-salt (○) and Mg (●) buffers in 2.2% gels for about 2 hrs. The mobilities of *Amoeba* RNA and BMV RNA were determined relative to those of HeLa-cell-r-RNA and *E. coli* r-RNA in separate experiments and are normalized to fit on the same scale. ○ and ▽, Low-salt buffer; ●, ▼ and ▲, Mg buffer.

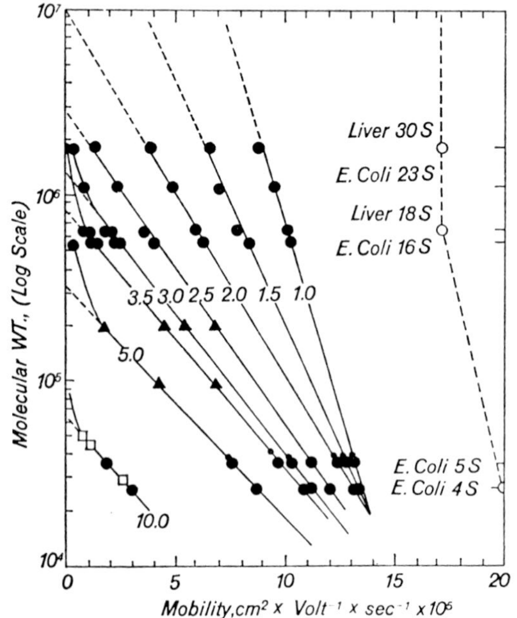

Fig. 13. Relation between the molecular weight of RNA species in composite agarose-polyacrylamidegels of various concentrations and their mobilites. From: Peacock and Dingman [P 9].

The gel concentrations (T % at C = 5%) are shown on each line. Agarose concentration: 0.5% throughout. The dotted line represents the mobilities in the free solution [O 4]. ▲, □ and small circles represent unknown cytoplasmic and nuclear RNAs used to determine their MWs.

large increase in S, yet almost no increase in mobility, indicating an opposing action of Mg^{2+} on size and charge of RNA such as masking of internal charges. The loss of charge is less effective in larger molecules than in smaller, as demonstrated by the discontinuity in Fig. 12. This points to an interesting application of the method by which the extent of folding relative to that in other buffers and other conformational changes may be detected.

In summary, polyacrylamide gel electrophoresis is a rapid and accurate method to determine the relative size and MWs of macromolecular ions such as proteins and nucleic acids. The method offers several advantages over classical procedures of MW determination. However, careful standardization of techniques is essential [K 19a]: not only should the conditions of electrophoresis be standardized, but also those of gel formation (see page 54).

1.1.5. Discrimination of Proteins Similar or Different in Charge and/or Size by Polyacrylamide Gel Electrophoresis

As outlined on page 4, gel electrophoresis separates proteins according to both charge and size. Often the question may be raised, whether proteins differ only in charge or, in addition, also in size. When different in size, one may ask whether, under defined experimental conditions, polymers or subunits of a basic protein unit do occur. Which methods are available to answer these questions?

According to eqn. (2) and (3) on page 9 (Fig. 5), the slope of log mobility versus MW plots is proportional to molecular size and, according to eqn. (2) on page 9, the ordinate intercepts of such lines are equal to log free mobilities m_0 (or 0 % gel densities) where m_0 is related to charge by eqn. (2) on p.: 20 $m_0 = Q/6 \pi r \eta$. It follows that (1) *parallel lines* in such plots would

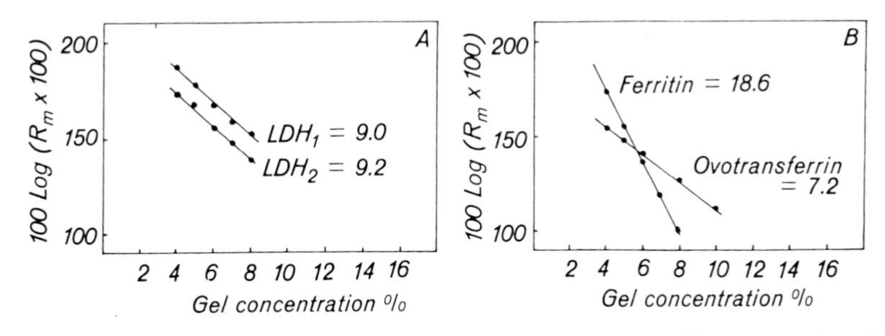

Fig. 14. The effect of different gel concentrations on the mobility of proteins (A) similar in size but different in charge (lactate dehydrogenase isozymes) and (B) different in size *and* charge. From: HEDRICK and SMITH [H 15].

R_m = distance of migration relative to the distance of the bromophenolblue front. Gel concentration refers to acrylamide content (constant Bis/acrylamide ratio of 1 : 30, w/w). The negative slopes of the lines are noted (see page 9).

indicate the occurrence of proteins of *identical size but different charge* (example: isozymes, Fig. 14 A); (2) *non-parallel lines*, however, would point to proteins of *different size* (Fig. 14 B); and (3) *non-parallel lines with slopes differing from each other in a similar degree*, thus revealing a regular family of curves (Fig. 4), would signify the existence of *polymers* or, conversely, of *subunits* of a basic protein unit.

As pointed out by HEDRICK and SMITH [H 15], important conclusions may be drawn from such plots. The possibility to distinguish between size and/or charge differences in proteins would allow, for example, to decide which the next step should be in a purification process of proteins (separation method resolving on the basis of size: gel filtration, or on the basis of charge: ion exchange chromatography). Furthermore the purity of a protein preparation may be tested with more confidence. For the possibility exists that a large protein, highly charged, and a relatively small protein, only slightly charged, will migrate at the same rate. *The appearance of a single band on disc electrophoresis should, therefore, not be interpreted as unequivocal evidence of protein homogeneity.* This may be concluded only if a single protein band is observed at several different gel concentrations.

1.2. Important Principles of Electrophoresis

When a particle of effective electrical charge Q is forced to migrate in a viscous medium (liquid or gel) by action of an electrical field (potential gradient) E this process is generally defined as electrophoresis. The driving force, which acts on the particle migrating with constant velocity, is equal to the frictional resistance f which the particle must overcome in the medium:

$$QE = f \qquad (1)$$

In free solution, the frictional resistance obeys Stoke's law: $f = 6\pi r v \eta$, where r represents the particle radius, v its velocity and η the medium viscosity. In gels, the frictional resistance is a yet undetermined complex function of gel density and particle size among others.

The *electrophoretic mobility* m of a particle is defined by:

$$m = \frac{d}{tE} = \frac{v}{E} = \frac{Q}{f} \qquad \left[\frac{cm^2}{Volt \cdot sec}\right] \qquad (2),$$

where d represents the migration distance of the particle in time t.

The particle mobility is a physical constant under defined electrophoretic conditions. It follows from eqn. (2) that the absolute mobility is obtained by dividing the measured migration distance by the time of electrophoresis and the applied potential gradient. Provided that different experiments are performed under identical conditions (t and E), it is permitted to compare migration distances directly. Thus, several authors expressed their results as relative mobilities. However it should be kept in mind that the migration distances are not equal but only proportional to the mobilities.

Since the *potential gradient* or *field strength* E corresponds to the ratio of current density J ($A \cdot cm^2$) to the specific conductivity $\varkappa$ ($\Omega^{-1} \cdot cm^{-1}$),

$$E = \frac{J}{\varkappa} \quad \left[\frac{\text{Volt}}{\text{cm}}\right] \tag{3},$$

the *velocity* v of the particle is given by:

$$v = Em = \frac{mJ}{\varkappa} \quad \left[\frac{\text{cm}}{\text{sec}}\right] \tag{4}.$$

Moreover, other factors influence the particle mobility. Thus, the *buffer pH* determines the net charge of the ion depending on the dissociation characteristics of the weak acid, base or amphoteric compound. The *ionic strength* determines the electrokinetic potential which reduces the net charge to the effective charge. At near approximation, the mobility is inversely proportional to the square root of the ionic strength [G 38].

The *ionic strength* μ is given by:

$$\mu = \frac{1}{2} \sum_{i=1}^{i=n} c_i z_i^2 \tag{5}$$

where c_i = molar concentration of the single ionic species; z_i = charge (valence) of the ionic species i in a solution containing n ions.

Low ionic strengths of buffer solutions allow high migration velocities with little heat evolution, while high ionic strengths lead to low migration velocities and increased heat generation, although the zones become sharper.

The electrical current to be applied has its limit by the *ohmic (Joule) heat* generated in the gel. Ionic mobility and free diffusion increase with increasing temperature, while medium viscosity decreases. Mobility increases by about 2.4 % per degree rise of temperature [M 67]. The *heat H* generated by the current I per unit of time is described by:

$$H = \frac{R \cdot I^2}{A} = \frac{V \cdot I}{A} = \frac{V^2}{R \cdot A} \quad [\text{Watts}] \tag{6},$$

where R = electrical resistance of the medium, V = voltage, and A = mechanical equivalent of heat, 4.185×10^7 erg/cal.

Dissipation of ohmic heat is a major problem for each electrophoretic method. Cooling of any electrophoretic medium gives rise to a temperature gradient between the inner and outer medium layers depending on the thickness of the medium. According to the form of the gradient this may lead to more or less distorted (e. g. curved) bands of the separated molecules. The conditions permissive for any electrophoresis can be deduced from eqn. (6). It should be noted however, that resistance may vary during electrophoresis (see page 40). The resulting changes of heat output at constant current or constant voltage can be calculated from eqn. (6) as well.

It is apparent from the equations and the discussion that electrophoretic migration depends on a number of *parameters*, such as size, shape, concentration, electrical charge, degree of hydration and dissociation of the particle; viscosity, pH-value, temperature and ionic strength of the medium; and electrical field strength (voltage gradient) and migration time.

The relationships between electrophoretic mobility, migration velocity, field strength, pH value and ionic strength are equally valid for both free flow and supported electrophoresis. In supported electrophoresis additional factors may become manifest which influence the mobility and sharpness of separation to a variable extent: adsorption, inhomogeneity and exchange capacity of the support material, electroosmosis, suction effects, heating and evaporation influences. Polyacrylamide gel has the advantage that adsorption and electroosmosis do not play a role [K 12, R 14, R 15, W 19].

1.3. Theory of Disc Electrophoresis: Polyacrylamide Gel Electrophoresis with Discontinuous Voltage and PH Gradients (Ornstein-Davis-System)

1.3.1. Introduction and General Comments

In disc electrophoresis [O 9] we are dealing with a *discontinuous separating system* with regard to pH value, buffer composition and gel pore size in which polyacrylamide gel serves as the matrix.

While continuous polyacrylamide gel electrophoresis (RAYMOND [R 16]) makes use of homogeneous buffer systems of uniform pH values, disc electrophoresis introduces buffers of different pH values and composition as well as gel layers of different pore size to create discontinuous voltage and pH gradients (ORNSTEIN and DAVIS [O 9, D 7]). Prior to separation, these *discontinuities* produce very thin, i. e. highly concentrated, starting zones which determine the sharpness of the separations.

Three physical effects are responsible for the high resolving power of disc electrophoresis:

1. The *concentrating effect* according to KOHLRAUSCH's regulating function [K 27], by which the sample components are concentrated in large-pore spacer (stacking) gels to form sharply defined zones prior to their separation.

2. The *molecular sieving effect* [S 58] by which the individual molecules are electrophoretically separated on the basis of their size (molecular weight) and shape or tertiary structure (see p. 4 for more details).

3. The *electrocharge effect* by which the sample molecules are fractionated according to their net electric charge.

1.3.1.1. Concentrating Effect

The "*regulating function*" of KOHLRAUSCH [K 27] as an expression of the stationary process means that during an electrophoretic separation, concentration shifts of ionic species can take place in every part of the medium only in such a manner that the sum all ratios of concentration c and mobility m of the components (cations B^+ and A^-) remains constant [A 30]:

$$\frac{c_{B\oplus}}{m_{B\oplus}} + \frac{c_{A\ominus}}{m_{A\ominus}} = F \quad (1)$$

The practical consequences of this function for the concentration of sample solutions are demonstrated by an example on p. 27; it also shows the necessary experimental conditions to produce the concentrating effect. At this point, we will briefly explain its theoretical principles. For further details concerning these interrelationships see, for example [A 30, E 2, L 39, M 67, O 9, W 1].

Let a solution of protein P⊖ in a buffer solution with ions B⊕ and G⊖ (glycinate) be applied on a buffer solution with ions B⊕ (base cations) and Cl⊖ (chloride as acid anion); the protein solution finally is covered with the latter buffer solution [M 67] (Fig. 15 A). A sharp and stable *moving boundary wx* forms between glycinate and chloride ions, provided their concentration ratio obeys the equation:

$$\frac{G}{Cl} = \frac{c_G \alpha_{Cl}}{c_{Cl} \alpha_G} = \frac{m_G\, z_{Cl}\, (m_{Cl} - m_B)}{m_{Cl}\, z_G\, (m_G - m_B)} \quad (2)$$

where m = mobility, z = charge, c = concentration, α = degree of dissociation of the ionic constituents. In order to achieve concentration, the pH value in zones 1 and 2 is changed such that the effective mobilities of the anions take the following order: $m_{Cl}\alpha_{Cl} > m_P\alpha_P > m_G\alpha_G$. The pH value necessary to achieve this end is calculated by:

$$\text{pH} = \text{pK}_{\alpha(G)} + \log \frac{\alpha_G}{1-\alpha_G} \quad (3)$$

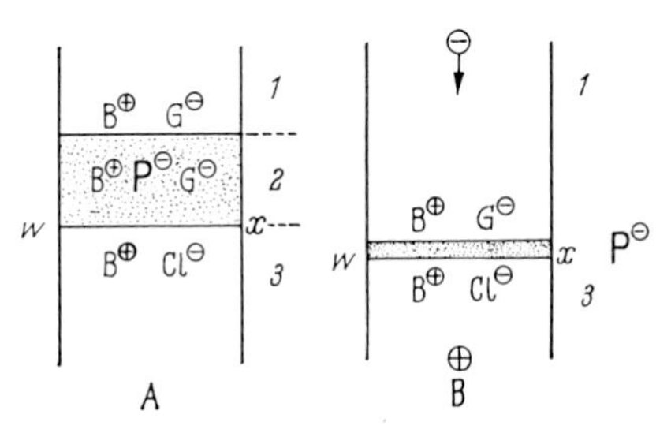

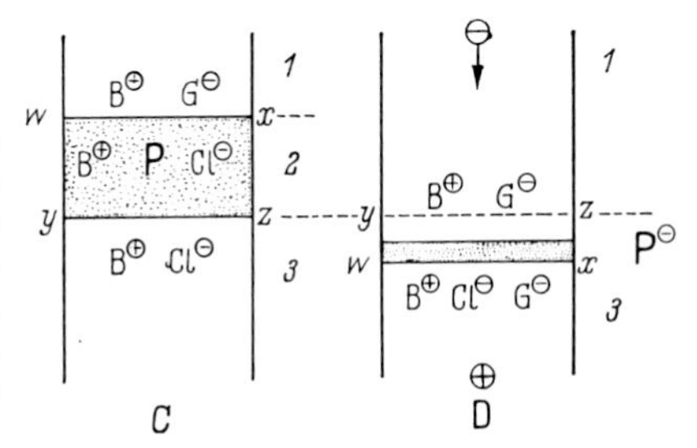

Fig. 15. - The theory of the concentration effect according to the Kohlrausch continuity or regulating function. From: MORRIS and MORRIS [M 67].

Concentration of a protein solution by mobility differences of the participating ions. Method I (A–B): proteins dissolved in upper buffer; method II (C–D): proteins dissolved in lower buffer.

Under these conditions, a progressive increase of conductivities takes place in zones 1, 2 and 3. With a flow of current, the proteins originally present in zone 2 collect in a thin layer on the boundary *wx* which now migrates down itself (Fig. 15B). At the same time, the protein concentration in this migrating boundary increases until the ratio c_p/c_{Cl} has reached a value which is determined by a relation similar to eqn. (2), differing from it only by the substitution of the G- with the P-values.

Another arrangement of the ionic species used in disc electrophoresis according to Ornstein [O 9] to produce a *stationary boundary* is shown in Fig. 15C. With a current flow, boundary *wx* moves down and concentrates the proteins in a thin layer as in the first case (Fig. 15D). The advantage of this arrangement resides in the possibility of maintaining a stationary front *yz* by selecting suitable concentrations of the base cations; consequently, a different pH value can be adjusted in zone 3 (separation gel) than in zone 2 (spacer gel). For example, if the pH value in zone 3 is raised to 9.5 – a value near the pK value of glycine – the effective mobility of the glycinate ions $m_G\alpha_G$ increases to such an extent that it exceeds that of the proteins: the glycinate ions therefore overtake the proteins which now migrate in a constant ionic medium.

While this system by Ornstein and Davis [O 9] is characterized by both voltage and pH discontinuities, the system by Allen [A 18, A 22] only utilizes voltage discontinuities to produce a moving boundary between two different anions for zone sharpening (see p. 29).

1.3.1.2. Advantages and Disadvantages of Continuous and Discontinuous Buffer and pH Systems

In 1957, Poulik [P 26] pointed out that the resolving power of starch gel electrophoresis could be significantly improved with *discontinuous buffer systems*. He used a borate buffer as the electrode medium, but a citrate buffer in the starch gel; he observed a boundary moving to the anode and attributed this to a *discontinuity in voltage gradient*. In polyacrylamide gel electrophoresis, Zwisler and Biel [Z 10] also found that a discontinuous Tris-borate system was superior to other buffer systems.

However, *borate* may complex with several sugars and lead to bent protein zones in gel electrophoresis. As shown by Lerch and Stegemann [L 18], plant extracts may contain such sugars creating band spreading. Moreover, bovine serum albumin interacts reversibly with borate buffers to give two or three bands in zone electrophoresis [C 8] which are not indicative of heterogeneity.

Anyhow, the concentration capacity of *disc electrophoresis* using discontinuous buffer *and* pH systems is unmatched until now. Thus, this method is particularly useful if very dilute samples and samples which cannot stand lengthy concentrating procedures are to be fractionated. In addition the risk of denaturating and degradation during such procedures is reduced.

The basic question of whether discontinuous buffer systems offer advantages over continuous systems has not received a consistent answer, however. Raymond [R 11] and Hjertén et al. [H 27] reject discontinuous systems, because in their view, sharp starting zones can also be obtained in *continuous systems*. Hjertén et al. [H 27] noted that some biological substances and structures (e.g. ribosomes) are stable only in certain buffer solutions within narrow pH ranges. Moreover, the high concentration of proteins in the stacking phase of disc electrophoresis may give rise to concentration artifacts (e.g. irreversible protein-protein interactions). This was also suggested for RNA molecules [M 40]. In all such cases, discontinuous buffer and pH systems undoubtedly are inappropriate. *For these reasons, attention should always be given to pH influences to which the sample is exposed in every disc electrophoretic run.* For example, the pH range in the standard alkaline gel system No. 1 (p. 44-45) extends between 6.7 and 9.3. In order to achieve similarly good separations with continuous systems, the starting zone must be maintained at a sufficient sharpness, i.e. at a high concentration, in these cases, too. Concentration is achieved by dissolving the sample in or dialyzing it against a buffer solution having a *lower ionic strength* – and thus lower conductivity – than that of the gel and electrode buffer solution (1 : 5 dilution, for example) [H 27].

When this technique is used to sharpen the zones, a low current intensity should be applied at the start of migration of the sample molecules into the separating gel in order to minimize the risk of thermal convection resulting from high electrical resistance. Subsequently, the optimum current intensity must be adjusted. However this method has a limited capacity, since many proteins require an ionic environment and a certain ionic strength to be soluble. Too much dilution may lead to aggregation and adsorption effects.

Since the separation effect is determined by the concentration of the sample available for any electrophoretic fractionation, one has to select a concentration method for each case: whether the sample needs to be concentrated before application to the gel, whether concentration can be obtained in a continuous pH system by the "dilution" method described above, or whether discontinuous buffer systems offer optimum separating conditions. Thus, the choice of any gel or buffer system must depend on the particular properties of the sample components to be separated as well as on the resolution required. Possible interferences during electrophoresis (e.g. protein-protein interactions, persulfate artifacts) must also be taken into account. In conclusion, there exists no universal gel or buffer system applicable to all possible separation problems. Both continuous and discontinuous systems offer advantages and drawbacks. Possible interferences resulting from pH discontinuities may be avoided by the use of a pH-continuous but buffer discontinuous system capable to achieve zone sharpening (see page 29).

1.3.2. Processes during Disc Electrophoresis

1.3.2.1. Concentration of the Sample Solution

Disc electrophoresis is carried out with small columns of polyacrylamide gel consisting of three parts (Fig. 16):

1. A *large-pore sample and anticonvection gel* containing the sample solution; in some cases, a sucrose solution may substitute it (see page 58).

2. A *large-pore spacer or stacking gel* in which the sample constituents are concentrated; and

3. A *small-pore separation or running gel* in which the sample constituents are separated.

The two ionic species which migrate in the running direction with the sample molecules, e. g. proteins, during electrophoresis and produce the discontinuous voltage gradient will be called *leading* and *trailing* ions. At the start of electrophoresis the leading ions are located in all three parts of the gel, while the trailing ions are present only in the buffer solution of the electrode

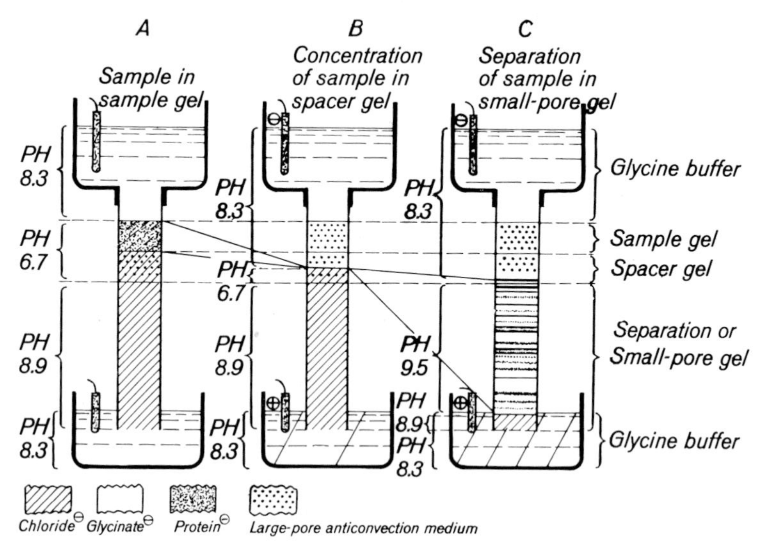

Fig. 16. – Theory of disc electrophoresis (explanation in text). From: ORNSTEIN [O 9].

reservoirs. By selecting a suitable pH value for the sample and spacer gel (p. 23), the effective mobilities $m\alpha$ of all ionic species should arrange in the following sequence:

$$m\alpha_{\text{leading ions}} > m\alpha_{\text{proteins}} > m\alpha_{\text{trailing ions}}$$

where leading ions, protein and trailing ions have the same charge, i.e. migrate to the same electrode. If the entire gel system is subjected to a flow of electrical current, the leading ions precede the proteins and trailing ions because of their higher mobility leaving behind a zone of lower conductivity (Fig. 17). According to eqn. (3) on p. 21, the specific conductivity is inversely proportional to the field strength (voltage gradient). This zone therefore attains a higher voltage gradient which accelerates the proteins and trailing ions to such an extent that they migrate behind the leading ions at the same velocity and consequently will stack up. According to eqn. (4) on p. 21, all ionic species move at the same velocity when the products of voltage gra-

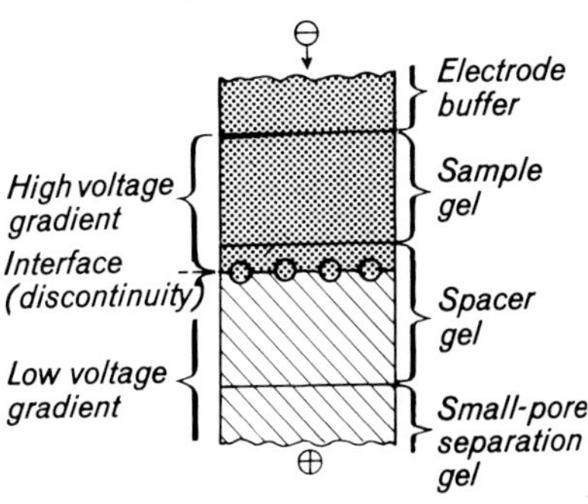

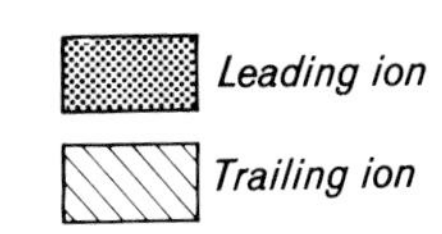

Fig. 17. – Theory of disc electrophoresis (explanation of the concentrating effect; see Fig. 16 B; other explanations in the text). From: WILLIAMS and REISFELD [W 20].

dient and mobility are equal to each other: $v = E_L \cdot m_L = E_T \cdot m_T$; v = velocity, E = field strength (voltage gradient), m = mobility, indices L = leading ions, T = trailing ions. A steady state is set up in which the products remain equal. Consequently, an interface forms between the leading and trailing ions which at the same time represents the front between zones of high and low voltage gradient; it can be observed in a practical experiment in the form of a descending schlieren front. This boundary now moves rapidly through the sample and the spacer gels. Since the mobilities of the sample proteins are intermediate between those of the leading and the trailing ions, they collect in the moving boundary and stack up as a narrow, highly concentrated protein zone. Individual proteins thus form a sequential arrangement, stacked into discs of micron-thickness in order of their mobilities and located between the leading and trailing ions. According to the KOHLRAUSCH regulating function [K 27], the final concentration of this protein band is independent of the initial concentration but is proportional to the leading ion concentration.

ORNSTEIN [O 9] has illustrated this concentrating effect by a practical example. For this purpose, he developed eqn. (2) (p. 23) which represents a modified expression of the KOHLRAUSCH regulating function (p. 23). If we let G = glycinate⊖, B = K⊕ and Cl = Cl⊖, where $m_G = 15$, $m_B = +37$ and $m_{Cl} = -37$ mobility units (10^{-5} cm²/V · sec), then eqn. (2) permits us to calculate glycinate/chloride = 0.58. Above pH 8.0, serum proteins have mobilities of between -0.06 to -7.5 units. According to theory, the trailing ion must have a lower effective mobility $m_G\alpha_G$ than the slowest individual protein fraction (e.g. -0.5). With a degree of dissociation $\alpha_G = 0.5/15 = 1/30$, this condition would be satisfied. As calculated from eqn. (3) (p. 23), a pH value of 8.3 is required for glycinate to achieve this degree of dissociation ($p_{K_2}^{25°C} = 9.778$ [NH_3^+]). For albumin molecules (molecular weight = $6.8 \cdot 10^4$, $m_P \cong -6.0$ mobility units, $z_P \cong -30$), which are to be concentrated in this glycinate solution, the ratio of albumin/chloride = $9.3 \cdot 10^{-3}$ (P = G) according to eqn. (2). In an 0.06 M chloride solution the albumin solution automatically concentrates to $5.6 \cdot 10^{-5}$ M (3.8 %) independent of the concentration of the original albumin solution, the length and the diameter of the gel column. With an initial 0.01 % albumin solution, for example, a 380-fold concentration would be obtained. – With the use of coloured proteins, e. g. phycocyanine, the effect can be directly observed.

1.3.2.2. Separation of Sample Components

When the migrating protein zone, sandwiched between leading and trailing ions, reaches the interface between spacer and separation gel, both the pH value and the gel density (pore size) change abruptly. The "new" pH value in the smallpore gel is selected such that the degree of dissociation of the trailing ions – and thus their mobility – will increase by a multiple. As a result, the trailing ions attain a mobility nearly equal to that of the leading ions. The trailing ions now overtake all proteins and precede the proteins directly behind the leading ions. This manoeuvre also changes the voltage gradient such that the proteins will now move in a zone of uniform voltage gradient and pH value under conditions similar to ordinary zone electrophoresis. The proteins are separated according to molecular weight, shape and charge by the suitably selected gel density. Thus, proteins of identical free mobilities but different molecular weights (different diffusion constants) will also migrate at notably different mobilities and consequently will be separated.

In our example, the "new" pH value to be selected for the small-pore gel is calculated in the manner described on p. 27. In a 7.5 % gel, the "fastest" serum protein, prealbumin, has a mobility of <-5.0 units. The effective mobility $m_G\alpha_G$ of the glycinate ion therefore must amount to at least –5.0 units, which is obtained by increasing the degree of dissociation (to $\alpha_G = 0.33$). According to eqn. (3) (p. 23), we thus obtain a pH value of 9.5.

To achieve electrical neutrality and sufficient buffering capacity of the entire gel system, cations or anions of weak bases and acids as counterions are selected such that (14 – pK_b) and (14 – pK_a), respectively, fall between the pH values of the spacer and the separation gel so that both gels will be buffered. Ornstein [O 9] as well as Richards et al. [R 34] have offered equations to calculate the base and acid concentrations for large-pore as well as for small-pore gels which are necessary to develop and maintain the calculated pH values even through prolonged runs and under higher electrical currents.

1.3.3. Guidelines for the Choice of Suitable Buffer and pH Systems

Williams and Reisfeld [W 20] deserve credit for having developed and compiled the general guidelines for the choice of buffer and pH suitable systems for disc electrophoresis of very diverse substances. Table 2 furnishes a guide for the preparation of a suitable system. It is a general prerequisite that the pH value and buffer composition of a system must afford a maximum resolution of the substances while maintaining their solubility and their chemical and biological stability. Furthermore, at the pH values of the sample, spacer and separation gels, the substances should carry the same charge as the leading and trailing ions. The pH value of the sample solution should not significantly change the pH value of the sample and the spacer gel. It is therefore advisable to check the pH value of the sample solution before it is applied and if necessary, to adjust it with a pH-meter to at least ± 0.75 pH units of the buffer.

For a practical application of Table 2, the components of the system satisfying the set conditions are selected in steps (from 1 to 6). Examples are given for cationic and for anionic systems; the left column lists respective theoretical values which for a comparison are faced by others in the right column, independent of those selected for practical purposes in this guide. See [W 20] for a detailed discussion of the range of application of these guidelines.

Richards et al. [R 34] have formulated equations permitting a calculation of the concentrations of the leading, trailing and counterions, the pH values of the spacer and separation gels as well as of the electrode buffer after selection of a weak, monovalent acid and base of suitable pK value. The method furnishes not only the desired separation pH but also a sufficient buffering capacity.

1.4. Polyacrylamide Gel Electrophoresis with Discontinuous Voltage Gradients (Allen-System)

Thin starting zones of the sample materials to be separated are an essential prerequisite for a high electrophoretic resolution . Besides the concentration method by Ornstein using both voltage *and* pH discontinuities to this end (see page 22), concentration may also be achieved by decreasing the conductivity of the sample with respect to the separating medium [H 27] (see page 25). This method avoids possible harmful effects arising from pH discontinuities, but has a limited concentrating capacity. Poulick [P 26] has demonstrated that, even during the run, already separated protein zones can be additionally sharpened by a discontinuous voltage gradient.

Allen [A 18, A 22] combined these methods (*concentration by a conductivity shift* and *zone sharpening by a moving boundary*) to set up a two-phase vertical slab system at a uniform pH throughout. The sample is placed in buffered

Table 2: Guide for the preparation of pH and buffer systems for disc electrophoresis. Conditions for the individual components of the systems. From: WILLIAMS and REISFELD [W 20].

Component of the system	Conditions and remarks on the choice of components for Anionic systems | Cationic systems	Examples for Anionic systems [W 20]	[D7,0 9]	Examples for Cationic systems [W 20]	[R 23]
1. pH value of separation gel	During electrophoresis, elevation | reduction of pH value by 0.5 units. Development of high degree of dissociation of trailing ion (up to 50%).	8.9 9.4	8.9 9.4	4.3 3.8	4.3 3.8
2. Leading ion	Higher mobility than that of all other ions. Independence of pH value.	Cl⊖	Cl⊖	K⊕	K⊕
3. Trailing ion (pK_a)	Weak acid | Weak base or amino acid with pK_a higher | lower by 1 pH unit than that of the separation gel. Lower mobility in spacer gel but higher mobility than that of all substances in the separation gel.	8.9–9.9	9.8 (Gly⊖)	3.3–3.4	3.6 (β–Ala⊕)
4. pH value of sample and spacer gel	pH value by 2–3 units below | above the pK_a of the trailing ion. Development of a low degree of dissociation of the trailing ion (1 to 0.1%).	6.8–7.8	6.7	5.6–6.6	6.8 (too high)
5. Buffer of separation gel (pK_a)	Weak base | Weak acid with pK_a up to 1 pH unit lower | higher than that of the separating gel. Availability of counter-ions to preserve electrical neutrality. Optimum buffer action in the pH range of the separation gel.	7.9–8.9	8.1 (Tris⊕)	4.3–5.3	4.8 (acetate⊖)
6. pH value of electrode buffer	Identity with pK_a value of separation gel buffer.	8.1	8.3	4.8	4.5

Abbreviations: Gly⊖ = glycinate, β–Ala⊕ = β–alanine, Tris⊕ = tris(hydroxymethyl)aminomethane.

sucrose solution of lower conductivity than the adjacent separation gel. A cap gel, 1.5 – 2 cm high depending on sample column height, is cast on top of the sample. Its buffer composition equals that of the sample solution. Its purpose is to prevent the moving boundary between leading and trailing ions overrunning the sample prior to accomplishment of phase 1 (Fig. 18) and to provide a constant conductance over the entire cell width. Initially the leading ion is present in the gel and sample buffers, the trailing ion in the electrode buffer. Under electric current the sample components will be sandwiched as they enter the separation gel due to the conductivity shift. They are initially fractionated according to their charge and size at a continuous pH level (phase 1). The moving boundary, operating under conditions described by the KOHLRAUSCH regulating function (page 23), then reaches each separated sample component and sharpens each zone due to the conductivity jump (or discontinuity in voltage gradient) which accelerates the zone rear rather than the zone front (phase 2). After passage of the boundary through each zone the separation proceeds, as in ordinary zone electrophoresis, at an increased pH of approximately 0.3 to 0.5 units higher than the initial pH due to the KOHLRAUSCH function. This pH increase depends on the buffer ions chosen.

Since no pH discontinuities are required to program the mobility of the trailing ion, this system offers a wide choice of buffer ions and pH conditions suitable for the separation of most biological materials. ALLEN [O 10] has developed several buffer systems for this purpose (Table 7, pages 50-51).

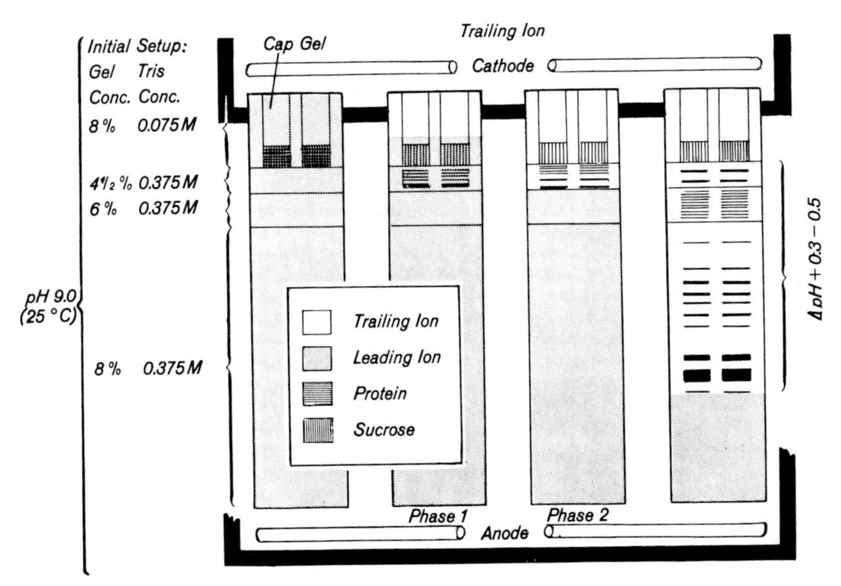

Fig. 18. Theory of polyacrylamide gel electrophoresis with discontinuous voltage gradients (Allen-system): Processes during performance. From: ALLEN [A 18].

2. METHODS OF ANALYTICAL DISC ELECTROPHORESIS

2.1. Physicochemical Properties of the Gel Components and their Identification

2.1.1. General Comments

The most important physical and chemical *properties* of the starting materials for the preparation of polyacrylamide gel are compiled in Table 3.

The following chemical reactions of *acrylamide* should be noted [A 28, C 10]:

Hydrolysis with the formation of acrylic acid and NH_3, highly favored by acid and alkali.

Reactions with hydroxyl compounds, such as phenols or aliphatic alcohols, lead to β-aryloxy- or alkoxypropionamide.

Reactions with NH_3 at 20° C lead to β, β', β''-nitrilotrispropionamide $N(CH_2$-CH_2-CO-$NH_2)_3$.

Reactions with mercaptans produce β-alkylmercaptopropionamides.

The polymerization process can also be initiated by ultrasonics, γ-radiation and natural light. If the hydrocarbon chains assume an adjacent position at a high degree of polymerization, the amide-groups will undergo intermolecular polycondensation with the formation of imide bridges.

The listed properties indicate that acrylamide and its aqueous solutions are best *stored* in brown bottles at temperatures of about + 4° C in the absence of light. Under such conditions, solid acrylamide is relatively stable, while the stability of aqueous solutions is limited to 1–2 months.

Similar conditions also apply to N,N'-methylene-bis-acrylamide (bis-acryloylamino-methane) and particularly to ammonium persulfate, $(NH_4)_2S_2O_8$. The solutions of the latter are stable only for about one week at 4° C. N,N,N',N'-tetramethylethylenediamine (1.2– bis [dimethylamino]-ethane) should also be stored in the absence of light.

For the use of other cross-linking comonomers (ethylene-diacrylate and N,N'-diallyl-tartardiamide) see page 60.

2.1.2. Concentration Determinations

– The concentration of organic substances can be determined by nitrogen analyses according to KJELDAHL. A potentiometric determination in acetanhydride by titration with perchloric acid in glacial acetic acid can also be used for acrylamide

[W 21]. The customary methods for the carbon double-bond determination, such as the hydrogenation value and iodine value, are of practical importance; other methods are described by LOGEMAN [L 38] and TROMMSDORFF [T 22].

Table 3: Properties of starting materials for the preparation of polyacrylamide gel.

Properties	Acrylamide	N,N′-methylene-bis-acrylamide	N,N,N′,N′-tetra-methylethylene-diamine
Formula	$CH_2{=}CH{-}CO{-}NH_2$	$CH_2{=}CH{-}CO{-}NH{-}CH_2{-}NH{-}CO{-}CH{=}CH_2$	$(CH_3)_2N{-}CH_2{-}CH_2{-}N(CH_3)_2$
Molecular weight	71.08	154.16	116.20
Appearance	White crystalline substance	White crystalline substance	Colorless to pale yellow oil
Odor	—	—	Ammoniacal
Solubility (g/ml at 30° C)	2.0 H_2O; 1.1 methanol; 0.4 acetone; 0.4 $CHCl_3$	0.31 H_2O (20° C)	Miscible with H_2O and organic solvents
M.p. or b.p. (°C)	84.5 ± 0.3	185 with polymerization	120–122
Density	1.122 [D_4^{30}]	—	0.777 [D_4^{20}]
Refractive index [n_4^{20}]	—	—	1.4184 ± 0.0005
Modification due to storage	Slow spontaneous polymerization	Slow spontaneous polymerization	Yellowing
Preparation	[C 10, S 24]	[F 17]	BEILSTEIN *4*, 250, 271 (415)

2.1.2.1. Determination of Acrylamide [A 28] and N,N′-methylene-bis-acrylamide

Reagents: 0.1 N $KBrO_3$-KBr (2.79 g $KBrO_3$ + 10.0 g KBr per liter), 6 N H_2SO_4, 20 % KI solution, 0.1 N $Na_2S_2O_3$, 1 % starch solution.

Procedure: The carefully weighed substance (1.0–1.5 g) is dissolved in dist. water to a volume of 500 ml. 0.1 N $KBrO_3$-KBr (25 ml), is mixed with 25 ml of the sample solution in a 250 ml flask for an iodine value analysis provided with a double-bore stopper and a separatory funnel and a glass connector (for vacuum application). 6 N H_2SO_4 (5 ml) is added dropwise under vacuum (about 12 mm torr). After standing for 20 min in darkness with frequent shaking, 15 ml KI solution is added. Upon relieving the vacuum, titration is carried out immediately with starch as indicator. A blank test is necessary.

Calculation: $\% \text{ substances} = \dfrac{(x-y)\cdot z}{\text{charge } (g)}$

Blank test: consumption of 0.1 N $Na_2S_2O_3$ = x ml
Experiment: consumption of 0.1 N $Na_2S_2O_3$ = y ml
Acrylamide: $z = 3.554$; methylene-bis-acrylamide: $z = 3.854$.

2.1.3. Purity and Purification Methods

The *purity* of *acrylamide* may differ considerably depending on its origin. On aging of aqueous solutions, the pH value increases as a result of hydrolysis; moreover, free acrylic acid causes a drop of the electrophoretic migration velocity [R 13]. In such cases, saturated solutions (about 15 g acrylamide/100 ml ethylacetate at 70° C), after filtering off the insoluble residue, are recrystallized. A second recrystallization from chloroform at 50° C (11 g/100 ml) in a similar manner completes the purification [A 23]. Later ALLEN [O 10] refined this procedure for producing a monomer which is both reproducible and suitable for use in quantitative studies on isoenzymes.

90 g of acrylamide are dissolved in a mixture of 300 ml ethylacetate and 300 ml of benzene, heated to + 70° C, filtered hot and recrystallized at + 20° C. The crystals are dried by vacuum filtration (BUCHNER funnel) until there is no further odor of benzene. They are redissolved in 540 ml of chloroform, heated to + 60° C, filtered hot, and cooled to no lower than + 22° C. Preferentially the crystals are stored under chloroform and dried by vacuum filtration when needed. The aqueous solution of the recrystallized substance shows a pH value of 4.9 – 5.2 and can be used as long as the pH value does not vary by 0.4 units.

According to SHEPHERD [S 40], acrylamide can also be recrystallized from a filtered acetone solution after addition of 3 vols. of benzene. LOENING [L 35] dissolves acrylamide in benzene at 70° C, filters the hot solution and slowly cools it to 0° C; the crystals are repeatedly washed with heptane and dried in vacuum. A product of even higher purity is obtained when 70 g acrylamide are dissolved in 1 l chloroform at + 50° C, filtered hot, allowed to crystallize at –20° C, and the crystals then briefly washed with cold chloroform or heptane and dried exhaustively in vacuo. PANYIM and CHALKLEY [P 3] recrystallize acrylamide from water at 0° C to reduce the incidence of gel fracture observed in 15 % gels longer than 7.5 cm.

However, repeated recrystallization does not completely free acrylamide from acrylic acid which forms easily, particularly by alkaline hydrolysis [A 28]. Acrylic acid may be removed by ion exchange chromatography, but care must be taken to prevent spontaneous polymerization.

N,N'-methylene-bis-acrylamide (*Bis*) is dissolved in acetone (10 g/l) at 40–50° C and filtered hot [L 35]. The solution is slowly cooled to –20° C, and the crystals are washed with cold acetone by centrifugation or filtration and dried exhaustively in vacuo.

The purity of *N,N,N',N'-tetramethylethylenediamine* (*TEMED*) can be tested by gas chromatography. Commercial products attain a purity of better than 99 %.

Impurities can cause pronounced swelling of the gels.

2.1.4. Toxicity of Acrylamide

Finally, reference should be made to the toxicity of acrylamide [A 23, F 44]. Even in 1 % solution, the substance may cause skin irritations. Contact with the skin and inhalation of acrylamide dust should therefore be avoided. Furthermore central nervous system disturbances have been reported. The acute lethal dose (LD_{50}) for mice is 170 mg/kg.

2.2. Equipment for Disc Electrophoresis

2.2.1. Gel Tubes and Tube Holders

Gel tubes are preferably made from low-alkali glass. Some workers have used Plexiglas or polycarbonate plastic, but this practice is not recommended because it tends to cause protein leaks along the poorly bonded interface between gel and plastic tube wall. The tubes usually have a length of 65–70 mm and inside diameter of 5 mm (o. d. about 7 mm). They must fit firmly into the rubber grommets of the electrode reservoir without leakage. The rims should be completely plane. Sharp edges are removed with emery paper or by careful fusion in the pilot of a Bunsen burner; care must be taken that the inside diameter will not be changed even by fractions of a millimeter. All tubes for an electrode reservoir should be of equal length and consist of identical material (glass).

Separation gels with dimensions of 35 × 5 mm can be prepared in these tubes. These dimensions were selected because they permit separation patterns of about 30 mm length which Davis [D 7] considered an optimum compromise for both concentration and separation. With shorter runs the sample bands will form in higher concentration; when the runs are longer, separation becomes better but the zones become broader due to diffusion. Very long separations would produce broad, blurred bands which would be difficult to analyze quantitatively. The larger the diameter of the gels, the more ohmic heating is generated and the longer will it be necessary to stain and to destain. In the opinion of Davis [D 7], *cylindrical gel columns* are of advantage compared to gel strips because more uniform thermal gradients develop during electrophoresis; consequently, higher field strengths and shorter running times can be selected (but see p. 37).

The glass tubes are *cleaned* by placing them into 10 % bichromate-sulfuric acid, thoroughly rinsing with distilled water and drying (if necessary by dipping into acetone and blowing air through them). Davis [D 7] recommends that after rinsing with distilled water, the tubes be dipped into an 0.5 % aqueous solution of Kodak Photo-Flo solution (a nonionic detergent) and dried.

To *fill* the tubes with the gel solution, they are closed on one end with tightly fitting rubber caps or by being pressed against a Parafilm – covered sponge rubber base. Some people prefer to have the rubber caps cemented in a row on a smooth piece of wood or plastic.

2.2.2. Apparatus and Accessoires

The *apparatus* consists primarily of upper and lower buffer reservoirs and a lid with the platinum electrodes mounted therein. The upper buffer reservoir is punched and silicone-rubber grommets are inserted into the holes to hold the gel tubes. Unused holes are closed with plastic rods or rubber stoppers. Regardless of the size and shape of the vessels, the electrodes should always have the same distance from each gel center. Moreover, the gel centers should be equidistant from each other to maintain a constant voltage drop between the electrodes and each gel [D 7]. The apparatus should permit cooling of the gel tubes by precooled (4° C) electrode buffer solution contained in the lower buffer reservoir. The construction of home made apparatus has been described in detail by several authors [for example, B 36, B 39, C 34, D 7, L 8, M 25, R 49]. A commercial unit of firm 5 is shown in Fig. 19.

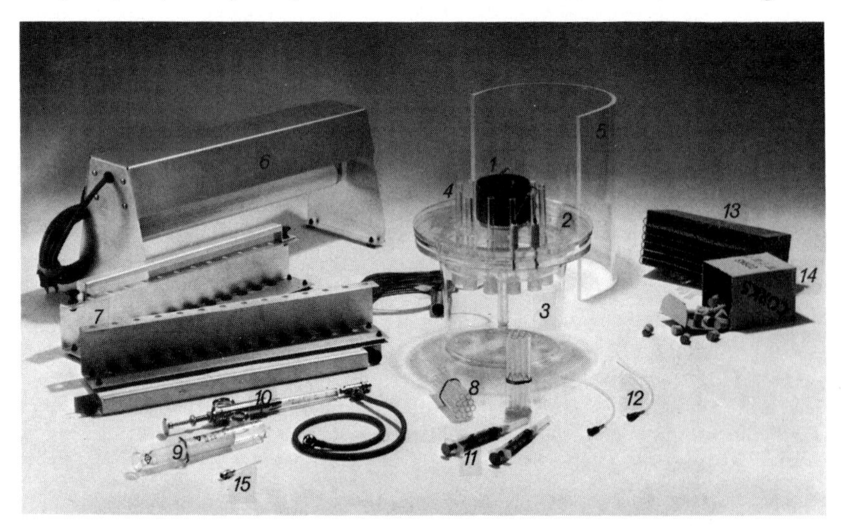

Fig. 19. Disc electrophoresis apparatus acc. to CANALCO (firm 5). (Power supply not shown).

1 : Electrode cable and lid, 2 : upper, 3 : lower electrode vessel (bath), 4 : upper bath stoppers, 5 : upper bath stand, 6 : preparation lamp, 7 : gel preparation racks and covers, 8 : gel tubes, 9 : 10 ml syringe, 10 : 1 ml automatic syringe, 11 : water layering syringes, 12 : syringe adapters and tubing, 13 : gel storage tubes, 14 : corks, 15 : gel removing needle.

The *additional equipment* comprises electrode cables, a suitable power supply (see page 40), racks upon which the gel tubes are mounted for loading, a polymerizing lamp, and various accessoires for handling the solutions and gels. The latter include microliter syringes or micropipettes, 1 ml and 10 ml syringes equipped with thin flexible tubings instead of needles and a stainless steel injection needle (about 8 cm long) with the tip blunted and filed smooth for gel removal. Equipment for staining and destaining, photographing the stained bands, and making microdensitometric traces of the patterns complete the required equipment.

The following companies are manufacturing apparatus for analytical disc electrophoresis: The firms 5, 6, 15, 26, 27, 31, 34. Of these, only CANALCO (firm 5) is licensed by the Mt. Sinai Hospital Research Foundation, New York, owner of the basic ORNSTEIN-DAVIS patent (US patent No. 3,384.564), to issue sublicenses for use of the disc electrophoretic technique in the United States of America.

Whereas all above apparatus use gels cast in tubular form, other investigators [A 20, A 37, N 18] have reported the development of moulds producing *gels of rectangular shape* (*strips*) which, in the authors opinion, may be conveniently photographed and subjected to quantitative densitometry without the claimed optical artifacts of cylindrical gels [A 37].

ARONSON and BORRIS [A 37] produce rectangular gel strips in glass cuvettes (manufacturer: Firm 33). To allow their use in the analytical apparatus, one end of the cuvette is fused into a rounded shape. However clean removal of intact rectangular gels may be difficult. Similarly ALLEN and MOORE [A 20] prefer Tiselius glass cuvettes, with bottom removed, and rectangular plastic tanks as electrode reservoir. NARAYAN [N 18] describes the use of splitcasting gel moulds (a two-piece block of Plexiglas with six channels) to permit easy and rapid removal of strips, after electrophoresis, for staining and destaining.

While in these methods each sample is electrophoresed in "its" rectangular gel, several authors find *vertical slab gel electrophoresis* of advantage because the patterns of samples electrophoresed under completely identical conditions in a *single* gel are thus directly comparable. This may facilitate pattern evaluation considerably, since the gels can be easily cut, the strips stained with different dyes to detect different molecular species (proteins, lipids, carbohydrates etc.) and scanned separately.

In the case of gel cylinders, it is sometimes observed that proteins migrate for different distances even though they are subjected to apparently identical electrophoretic conditions at the same time. This is caused by small differences in gel length, gel diameter or gel concentration. Different gel concentrations at the upper gel end may result from careless layering of water. However, more acrylamide is needed in slabs, and ohmic heating must usually be controlled by circulating coolant. For precise identification and comparison of separated material in gel cylinders, the split-gel technique may be applied (see page 64).

In principle, *vertical slab gel electrophoresis units*, such as that designed by RAYMOND [R 10] for continuous polyacrylamide gel electrophoresis, then further modified by many others [e.g. A 1, A 12, B 35, L 31, M 12, M 30, R 6, R 21, R 44, A 78, T 13, W 29] can also be loaded with discontinuous gel systems; it is only necessary to assure that the stacking gel can be photopolymerized. Fig. 20 presents, as an example, an apparatus distinguished by a gel block consisting of plates which can be cooled and separated for easy and rapid removal of the gel. Particularly, small *glass* cells made from micro-

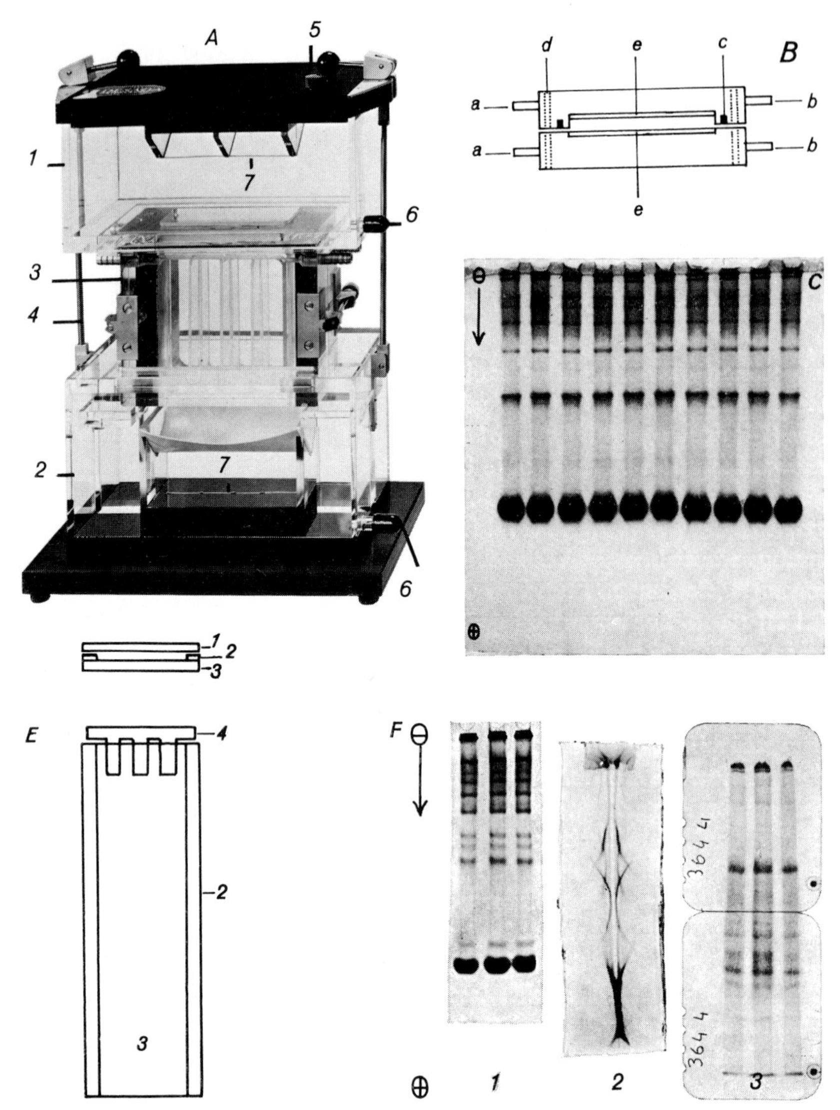

Fig. 20. Vertical macro and micro slab gel electrophoresis. From: Maurer [M 30, M 31a].

A: Diagram of the composed macro flat gel apparatus (available from firm 6), (1) upper, (2) lower electrode vessel, (3) gel block, (4) bow, (5) buffer inlet, (6) buffer outlet, (7) platinum electrode; the upper vessel has a slot surrounded by a rubber gasket to seal the gel block when tightened with the bow. B: Cross section of the gel block, (a) coolant inlet, (b) coolant outlet, (c) rubber seal, (d) bores for screws to clamp the block parts together, (e) glass plates. C: Human serum proteins separated by the macro apparatus using gel system No. 1a, gel size: 100 × 100 × 3.5 mm. Samples were applied to slots in the stacking gel by means of Teflon spacers. D: Cross section, E: top view of glass cuvette for micro flat gel electrophoresis, (1) cover microscope slide, (2) glass strip, (3) back slide, (4) Teflon comb. F: Micro flat gels, (1) IgM containing human serum proteins in 5.5 % gel, (2) immunodiffusion pattern of human serum proteins following electrophoresis and reaction with rabbit antihuman serum, (3) contact autoradiography of ^{14}C-labelled E. coli proteins by means of personal monitoring films exposed for 28 days.

scope slides can produce gel slabs of a size (75 x 18 x 0.7 mm) which allows a quick, simultaneous analysis of several samples (Fig. 20 D-F). This technique, developed by Maurer and Dati [M 31a], requires only about one fifteenth of the sample amount needed for ordinary disc electrophoresis (i.e. approximately 0.1 to 10 μg protein in < 1–3 μl solution), hence may be used on a microscale. Consequently, the time for staining and destaining is reduced considerably. The gel slabs can be easily scanned, dried and subjected to autoradiography and immunodiffusion. For other "slide" techniques see [A 24a, C 28a, G 11a].

Holmes [H 37] and Tombs [T 17] perform electrophoresis in horizontal gel plates (*horizontal slab gel electrophoresis*). This technique, however, is not generally recommended, since electro-decantation may lead to distortions in the electrophoretic patterns [A 12].

Tube or slab – it is apparently a matter of the separation task and of personal preference. Some arguments pro and contra have been discussed by Akroyd [A 12] and Gordon [G 23]. Akroyd [A 11] also observed a property of polyacrylamide worth to be mentioned: the gel adheres more firmly to glass than to Plexiglas. Thus, in molds of acrylic sheets, the sample may seep down the outer surface of the gel and give rise to tailing bands. This observation was confirmed by Stuyvesant [S 78] and Blattner [B 35] who also prefer glass plates for this reason. Nonionic detergents like Tween 80 (0.1 % in all gel solutions) [R 44], Triton X-100 [N 27] or Brij 35 [P 11] may provide a good wettability of the mould wall, but may also dissociate proteins (see page 67).

2.2.3. Electrophoretic Destaining Equipment

Essentially, the same equipment as for electrophoretic separation can be used. The gel tubes for destaining are 1 cm longer, however (i.e. about 8 cm in length) and have a larger inside diameter (6–7 mm) since the gels swell during fixing and staining as a function of the concentration of crosslinking agent. One end of the destaining tubes is tapered to a narrow orifice by fusing the rim, thus preventing the gels from sliding through. Destaining by electrophoresis may take place in the descending direction as in the case of electrophoretic separation, i.e. in the separation direction. However greater speed (10–15 min. versus 45–60 min.) is obtained by destaining the gels *perpendicular to the separation direction* (see also page 81). Suitable equipment has been designed by several investigators [F 8, M 25, M 28, O 11, P 14, P 30, S 22, S 27]. Commercial equipment embodying these principles is available e.g. from the firms 5, 6, 27, 31, 32.

Richards et al. [R 34] place the gels between two dialysis tubes filled with 10 % acetic acid (1 cm diameter) into which graphite electrodes and degassing tubes have been mounted. The destaining unit of Burger and Wardrip [B 60] or Nagy et al. [N 4] is suited for gel strips.

2.2.4. Power Supply Units

For *informative* analyses, unregulated power packs may be used which supply up to about 400 V and 5 mA d.c. per gel column. However, for *reproducible electrophoresis runs and for runs of long duration*, a constant current supply is generally used.

A run is only reproducible if the voltage drop is maintained constant in all parts of the support [D 18]. Power packs exist which supply constant regulated current at the outputs. During a run, the resistance may change in the electrophoretic medium and in the external circuit (on the electrodes, separators and in the electrode buffer solution), which again varies the part of the voltage gradient available for the separation. If constant voltage instruments are used (measured at the outputs), all resistance changes that occur in the support *and* in the outer circuit become noticeable, while the voltage drop with constant current instruments is independent of resistance changes in the external circuit. If the temperature of the separating medium changes, the migration speed changes more extensively in constant voltage than in constant current instruments according to BRATTSTEN [B 54]. For these reasons, constant current instruments are generally preferred.

In zone electrophoresis with paper, celluloseacetate, agar, starch, polyacrylamide, etc. as the supporting material, the migration velocity and its reproducibility are influenced by changes in temperature, ion composition and distribution more than in carrier free electrophoresis. This is particularly true for disc electrophoresis in which a distinct voltage increase is observed originating from the decreasing conductivity of the trailing ions moving through and out of the gel column. Consequently, the resistance increases while the current intensity drops. Without current regulation, longer running times are necessary, causing an undesiderable broadening of the sample bands due to diffusion.

For precisely reproducible electrophoresis runs it is therefore important to maintain a constant voltage gradient *in the entire separating field* (*supporting material*). DIKE and BEW [D 18] have designed a regulating power pack for this purpose. Most of the presently available instruments cannot guarantee such a constancy, however. The programmable power supply, by HOAGLAND [H 29], permits fixed control of the voltage applied to the electrophoresis cell as a function of time; thus a satisfactory reproducibility of the gel pattern may be obtained. For disc electrophoresis it is undoubtedly more suitable to use power supplies of constant current intensity than those without regulation; this affords a minimum running time which is directly proportional to the migration distance.

In any case, safety interlocks are desirable to cut off power to the electrodes when the electrophoresis bath is opened. Manufacturers of current regulated a–c power supplies are, for example, the firms 2, 4, 5, 12, 27, 28, 31.

Finally it should be pointed out that an increase of resistance is connected with an increase of *heat generation* according to eqn. (6) (p. 21). Cooling of the gel tubes by a precooled buffer solution and/or electrophoresis in a low-temperature laboratory at about + 4° C is useful for good results, particularly for heat labile samples (p. 102). Nevertheless, a *temperature gradient* between the interior and exterior of the gel can never be entirely avoided.

As discussed on page 21, the choice of any desired value of voltage or current is limited by the deleterious effects of ohmic heating due to the conversion of electrical energy (power) on passage through a resistance. Too low a power delivered to an electrophoretic cell will lead to poor separations, increased diffusion and long running times, whereas too high a power will result in excess of heat which, in turn, gives rise to convection currents, remix of separations, increased diffusion and destruction of labile samples. However, neither constant voltage nor constant current power supplies can provide optimum control of heating. ALLEN, MOORE and DILWORTH [A 22] solved this problem by constructing a *pulsed constant power supply* (manufacturer: Firm 22) which delivers constant output power to the electrophoretic cell, independent of the variations in resistance during the run. Thus, overheating is avoided; simultaneously high voltage polarization at low duty cycle should provide quick and sharp separations.

NEUHOFF et al. [N 29] designed a *power pack* for *micro-disc electrophoresis* which supplies voltage between 0 and 450 V and a constant current of up to 2 mA. The problem to regulate and maintain very low currents in the μA range was solved by the use of suitable noise diodes. Ten or more microelectrophoretic runs can be carried out simultaneously; their voltage and current values can be measured individually. Available from firm 29.

2.3. Procedures

2.3.1. Gel Systems and their Compositions

2.3.1.1. General Comments

The success of separations depends highly on the *purity* of the gel constituents. A purification of acrylamide and N,N-methylene-bis-acrylamide (Bis) can become worthwile; only ultrapure monomers lead to polymers which transmit UV radiation (λ = 260–280 nm) [L 35] (page 34).

Tables 4 through 7 list some customary gel systems for disc electrophoresis and other techniques of polyacrylamide gel electrophoresis. The chosen concentrations of gel components which are shown in the instructions for the solutions have proved to be suited for a number of analyses, but on the whole must be considered arbitrary. For special problems, an experimental determination of the optimum concentrations, particularly of monomer and comonomer, is therefore advisable. For the selection of suitable buffer and pH systems, the guide shown on p. 30 (Table 2) may be helpful. Instructions for gel systems for special separation problems are available; for details see the relevant chapters.

Gel systems Nos. 1 and 2 (Table 4) normally separate anionic substances with free mobilities in the range of –0.6 to –7.5 units. S_{α}–, S_{β}– and *γ-globulins*, which have intermediate mobilities (–0.75 to –1.0), can be resolved still better [J 14] if the pH of the spacer gel is raised from 6.9 to 7.3 and that of the separation gel is lowered from 9.5 to 8.6. In the gel system No. 1 a, the amount of Tris is replaced as follows: soln. No. 1: 10.65 g Tris, pH 7.2; soln. No. 4: 6.0 g Tris, pH 7.3; electrode buffer: 10.0 g Tris, pH 8.5.

For *basic proteins* (e. g. histones), SHEPHERD and GURLEY [S 40] recommend a buffer system (gel system No. 10) in which glycine is replaced by valine in the anodic (upper) buffer (compare gel system No. 7). Valine has a similar buffering capacity as glycine, but a lower mobility at pH 4.0. This leads to a steeper voltage gradient.

CHOULES and ZIMM [C 34] recommend 1 M propionic acid for *acid buffer systems;* difficulties in the polymerization of highly acid solutions are remedied by a persulfate-bisulfite catalyst system.

As mentioned on page 4, *urea* can be incorporated in the gels. In this case, all gel solutions are brought to the desired urea concentration (e.g. gel system No. 11); it is of advantage to increase the normality or molarity of the acids and bases with a suitable reduction of the ml-volume.

For *proteins* which are *unstable above pH* 8.5, not only gel system No. 6 but the following system can be recommended; it concentrates at pH 7.3 and separates at pH 8.5. In solution No. 1, 36.6 g Tris is replaced by 11.7 g imidazole and 0.23 ml TEMED is replaced by 0.24 ml TEMED (pH 7.9); in solution No. 4a, 5.98 g Tris is replaced by 2.93 g imidazole (pH 5.9); the electrode buffer solution contains 0.465 g imidazole and 14.4 g DL-asparagine in 1 l solution (pH 7.2–7.4) and is used in 10-fold aqueous dilution.

Peptides can be separated in a small-pore gel system which contains polyvinyl pyrrolidone (PVP). The following system might serve as a guideline. Sample gel: 2 parts soln. 4a, 4 parts No. 43 (Table 16, page 129), 2 parts No. 6, 1 part of a 5 % peptide solution, 7 parts 40 % PVP (K 30); "separation gel": 2 parts No. 4a, 4 parts No. 43, 2 parts persulfate soln. (0.56 g/100 ml), 1 part water and 7 parts 40 % PVP (K 30). After mixing check pH. Add a saturated solution of Tris dropwise to bring pH to 6.8 (initial pH will be lower). Polymerize the gels, allowing a minimum of an hour. Do not refrigerate solutions.

It is advisable to *store* all solutions in brown bottles at + 4° C (except for solution No. 11, Table 4). The pH value of solutions with a specified pH should be checked with a pH-meter and adjusted, if necessary. The solutions are stable for about 2–3 months except for all ammonium persulfate solutions (No. 9, 13, 18, 21, 26) and the saccharose solution No. 7 (watch for fungus growth!) which should not be stored for more than one week at + 4° C. Decomposition phenomena become manifest by a change of the pH of the solutions as well as by a retardation of the polymerizability.

2.3.1.2. Explanations to Table 4

– The designation "pH 8.9–7.5 %-gel" means that the separation gel of the corresponding system contains a buffer of pH 8.9 and 7.5 % acrylamide (w/v). "Concentration" and "separation" at pH 8.3 and 9.5, respectively, indicate at which pH values the substances are concentrated and separated during electrophoresis. All solutions should contain double-distilled water as the solvent. The polarity is applicable with the assumption that the sample substances migrate downward. Distilled water can be used in place of the saccharose solution (No. 7). The molecular weight values are approximate. All ammonium persulfate solutions must be added to the other solutions just before pouring of the gel. All solutions should be at room temperature (20° C) before mixing.

For the preparation of separation gels with acrylamide concentrations between 4 and 7.5 %, the equation $A = y \cdot 30/7.5$ is used for the calculation, where A = quantity of acrylamide in g which replaces the quantity of acrylamide in solution No. 2; y = desired gel concentration in per cent. – For gel concentrations above 7.5%, solution No. 2 is replaced by solution No. 8. In order to determine the volume parts B of solution No. 8 in the small-pore gel mixtures of gel system No. 1, the equation $B = y/7.5$ is used for the calculation. Moreover, for the determination of the volume parts C of water in the small-pore gel mixture, the equation $C = 4 - (B + 1)$ is used.

Example: Desired gel concentration: 15 %; B = 15/7.5 = 2 parts; C = 4 – (2 + 1) = 1 part. With a gel concentration of 22.5 %, no further water would need to be added. To prepare mixtures of still higher concentration, the concentration of solution No. 3 is increased: for example, from 140 mg ammonium persulfate/100 ml solution to 140 mg/75 ml. For a 30 % separating gel, 1 part of solution No. 1, 4 parts of solution No. 8 and 3 parts of solution No. 3a are needed (140 mg/75 ml). Concentrations above 30 % result in very rigid and brittle gels.

Table 4: Abbreviations

Tris	Tris(hydroxymethyl)aminomethane as base
Per	Ammonium persulfate, $(NH_4)_2S_2O_8$
SO_3	Sodium bisulfite, $NaHSO_3$
Ribofl.	Riboflavin
Sacch.	Saccharose (sucrose)
TEMED	N,N,N',N'-tetramethylethylenediamine
Bis	N,N'-methylene-bis-acrylamide
EDIA	Ethylenediacrylate
AcOH	Glacial acetic acid
MW	Molecular weight
p.	Volume parts

Table 4(a): Gel systems for continuous and discontinuous polyacrylamide gel electrophoresis.

Gel system		Range of application (examples)	Concentr. at pH	Separation at pH	Separation gel stock solutions			
No.	Designation				No.	Components per 100 ml solution	pH	Mixing ratio(v/v)
1	pH 8.9–7.5% (medium-pore gel)[1] [D 5]	Proteins of MW 10^4–10^6; optimum resolving range from $3 \cdot 10^4$–$3 \cdot 10^5$ (e. g. serum proteins)	8.3	9.5	1	1 N HCl 48.0 ml Tris 36.6 g TEMED 0.23 ml	8.9	1p. No.1 2p. No. 2 1p. H_2O 4p. No.3
					2[1])	Acrylamide 30.0 g Bis 0.8 g		
					3	Per 0.14 g		
2	pH 8.9–15% (small-pore gel)	Proteins of MW < $3 \cdot 10^4$	8.3	9.5	8	Acrylamide 60.0 g Bis 0.4 g		1p. No.1 2p. No.8 4p. No.3
3	pH 8.9–30% (small-pore gel)	Proteins of MW < 10^4, minimum about 2,000	8.3	9.5	9	Per 0.18 g		1p. No.1 4p. No.8 3p. No.9
4	pH 8.9-3.75% (large-pore gel)	Proteins of MW > 10^6, maximum about $2 \cdot 10^6$	8.3	9.5	10	Acrylamide 15.0 g Bis 0.4 g		1p. No.1 2p. No.10 1p. No. 6 4p. H_2O[3])
5	pH 7.3–7.5% (medium-pore gel) [T 1]	Basic proteins migrating back to cathode in gel system No. 1	8.3	6.6	11[4])	1 N KOH 8.0 ml Glycine 19.0 g TEMED 0.08 ml	7.3	6p. No.11 1p. No.12 1p. No.13
					12	Acrylamide 60.0 g Bis 1.6 g		
					13	Per 0.56 g		
6	pH 7.5–7.5% (medium-pore gel) [W 20]	Proteins (especially enzymes) showing optimum separation at pH 8.0 and lability above pH 8.0	7.0	8.0	15	1 N HCl 48.0 ml Tris 6.85 g TEMED 0.46 ml	7.5	1p. No.15 2p. No. 2 1p. H_2O 4p. No. 3
7	pH 4.3–15% (small-pore gel) [R 23]	Basic proteins (e.g. histones) of MW about $2 \cdot 10^4$	5.0	3.8	17	1 N KOH 48.0 ml AcOH 17.2 ml TEMED 4.0 ml	4.3	1p. No.17 2p. No. 8 1p. H_2O 4p. No.18
					18	Per 0.28 g		

[1]) For a 7% gel, solution No. 2 is replaced by No. 2a: 28.0 g acrylamide, 0.735 g Bis [D 7]

[3]) Mixture is photopolymerized.

[4]) Storage at +20°C.

Table 4(b) continued

Spacer gel stock solutions					**Electrode buffer solution**		Polarity	
No.	Components/100 ml solution		pH	Mixing ratio (v/v)	Components/1000 ml solution	pH	Top	Bottom
4[2])	1 M H_3PO_4 Tris	25.6 ml 5.7 g	6.9	1p. No.4 2p. No.5 1p. No.6 4p. No.7	Tris 6.0 g Glycine 28.8 g For use: 10% aq. soln.	8.3	—	+
5	Acrylamide Bis	10.0 g 2.5 g						
6	Riboflavin	4.0 mg						
7	Saccharose	40.0 g						
	As system No. 1			As system No. 1	As system No. 1	8.3	—	+
	As system No. 1			As system No. 1	As system No. 1	8.3	—	+
	As system No. 1			As system No. 1	As system No. 1	8.3	—	+
14	1 N KOH Glycine TEMED	48.0 ml 4.8 g 0.46 ml	10.3	1p. No.14 2p. No. 5 4p. No. 7 1p. No.13	Glycine 13.7 g 2,6-dimethylpyridine (2,6-lutidine) 38.2 ml For use: 10% aq. soln.	8.3	+	—
16	1 M H_3PO_4 Tris TEMED	39.0 ml 4.95 g 0.46 ml	5.5	1p. No.16 2p. No. 5 1p. No. 6 4p. No. 7	Diethylbarbituric acid 5.52 g Tris 1.0 g	7.0	—	+
19	1 N KOH AcOH TEMED	48.0 ml 2.87 ml 0.46 ml	6.7	1p. No.19 2p. No. 5 1p. No. 6 4p. H_2O	β-Alanine 31.2 g AcOH 8.0 ml For use: 10% aq. soln.	4.5	+	—

[2]) Soln. No. 4 can be replaced by No. 4a: About 48 ml 1 N HCl, 5.98 g Tris, 0.46 ml TEMED, pH 6.7 [D 7].

Table 4(a) continued

Gel system		Range of application (examples)	Concentr. at pH	Separation at pH	Separation gel stock solutions			
No.	Designation				No.	Components per 100 ml solution	pH	Mixing ratio(v/v)
8	pH 4.3–7.5% (medium-pore gel)	Basic proteins of MW > $2 \cdot 10^4$	5.0	3.8				1p. No.17 2p. No. 2 1p. H_2O 4p. No.18
9	pH 2.9–7.5% (medium-pore gel)	Highly basic proteins of MW > $2 \cdot 10^4$	4.0	2.3	20	1 N KOH 12.0 ml AcOH 53.25 ml TEMED 1.15 ml	2.9	4p. No.20 2p. No. 2 2p. No.21
					21	Per 2.8 g		
10	pH 2.9–20% (small-pore gel) [S 40]	Basic proteins (histones)			23	KOH 224.0 mg TEMED 0.4 ml AcOH to pH	2.9	5p. No.23 5p. No.24 5p. No.25 2p. No.26
					24	Acrylamide 60.0 g		
					25	Bis 1.2 g		
					26	Per 1.0 g		
11	pH 9.4–4% (large-pore gel) [R 25]	H- and L-polypeptide chains of γG-immunoglobulins (antibodies)			27	Acrylamide 16.0 g Bis 0.8 g 10 M urea[5]) to 100.0 ml		1p. No.27 1p. No.28 2p. No.29
					28	Tris 18.15 g 1 N HCl 24.0 ml TEMED 0.24 ml 10 M urea[5]) to 100.0 ml	9.4 at 25° C	
					29	Per 0.14 g 10 M urea[5]) to 100.0 ml		
12	pH 8.7–7.5% (medium-pore gel) [P 32]	Cytoplasmic brain proteins			33	Tris 4.48 g 1 N HCl to pH	8.7	2p. No.33 4p. No. 2
					34	TEMED 1.0 ml		1p. No.34 8p. H_2O 1p. No.35
					35	Per 1.2 g		
13	pH 6.5–7.5% (medium-pore gel) [C 34]	L-polypeptide chains of antibodies			37	NaOH 31.6 mg TEMED 0.115 ml Cacodylic acid 1.877 g	6.5	1p. No.37 1p. No.38 2p. No.39
					38	Acrylamide 30.0 g EDIA 2.0 g		
					39	Per 70.0 mg		

[5]) Deionize urea previously, e.g. by chromatography on a mix-bed ion exchanger; after this treatment, the conductivity should not exceed 5 μϰ.

Table 4(b) continued

No.	Spacer gel stock solutions: Components/ 100 ml solution	pH	Mixing ratio (v/v)	Electrode buffer solution: Components/ 1000 ml solution	pH	Polarity: Top	Polarity: Bottom
	As system No. 7		As syst. No. 7	As system No. 7	5.0	+	—
22	1 N KOH 48.0 ml AcOH 2.95 ml	5.9	1p. No.22 2p. No. 5 1p. No. 6 4p. H_2O	Glycine 28.1 g AcOH 3.06 ml For use: 10% aq. soln.	4.0	+	—
	Spacer gel solns. not used (preparation of separating gel soln. No. 23: dissolve KOH in 75 ml H_2O, add TEMED and adjust to pH 2.9 with AcOH, check pH after bringing to volume; yields 0.04 M potassium acetate.)			Upper buffer: 0.3 M DL-valine, pH adjusted with AcOH (pH 4.0) Lower buffer: 0.3 M DL-glycine, pH adjusted with AcOH	4.0	+	—
30	Acrylamide 8.0 g Bis 0.8 g 10 M urea[5]) to 100.0 ml		1p. No.30 1p. No.31 2p. No.32	Upper buffer: Tris 5.16 g Glycine 3.48 g 10 M urea 700.0 ml (pH 8.91) Lower buffer: Tris 14.5 g N HCl 60. 0 ml	8.07	—	+
31	Tris 2.23 g 1M H_3PO_4 12.8 ml TEMED 0.1 ml 10 M urea[5]) to 100.0 ml	6.74 at 25° C					
32	Ribofl. 1.0 mg Per 40.0 mg 10 M urea[5]) to 100.0 ml						
36	Tris 0.726 g 1 N HCl to pH	6.7	2p. No.36 4p. No. 5 1p. No.34 8p. H_2O 1p. No.35	Tris 6.0 g Glycine 28.8 g	8.3	—	+
	Spacer gel solns. not used (sample soln. should have lower ionic strength than soln. No. 37)			Solution No. 37	6.5	—	+

Table 4(a) continued

Gel system No.	Designation	Range of application (examples)	Concentr. at pH	Separation at pH	Separation gel stock solutions No.	Components per 100 ml solution	Mixing ratio (v/v)
14	pH 2.1–7.5% (medium-pore gel) [C 34]	Proteins showing optimum separations at pH 2.5 and associating at pH > 3	2.1	2.5	40	4 M propionic acid	1p. No.40
							1p. No.38
					41	Per 0.023 g	1p. No.41
					42	SO_3 0.035 g	1p. No.42
	Spacer gel solns. not used (pre-electrophoresis necessary). Sample soln. in 0.2 M propionic acid.		Electrode buffer: 1 M propionic acid pH 2.1				Polarity: Top + Bottom −

Table 5: Composition of buffer systems according to RICHARDS et al. [R 34, R 36].

Separation pH	Weak acid (A)	Weak Base (B)	Separation gel buffer soln. pH	Separation gel buffer soln. B(g)	Stacking gel buffer soln. pH	Stacking gel buffer soln. B(g)	Electrode buffer soln. pH	Electrode buffer soln. B(g)	Electrode buffer soln. A(g)
5.2	Acetic acid	Creatinine	4.8	11.4	2.6	5.68	3.8	0.41	1.8
7.2	Cacodylic acid	Imidazole	6.7	6.64	4.1	3.41	5.3	0.21	4.14
8.2	Diethyl-barbituric acid	Imidazole	7.8	5.66	5.8	3.42	7.0	0.23	5.52
8.9	Diethyl-barbituric acid	Tris	8.5	29.4	7.8	8.93	7.0	0.40	5.52
9.2	Glycylglycine	Ammediol[1]	8.0	12.6	6.1	5.25	7.0	0.108	3.96

To prepare both the separation and spacer gel buffer the indicated amount of weak base (B) is mixed with 50 ml 1 N HCl and water to give 1 l solution. To prepare the electrode buffer the indicated amounts of both base and acid are dissolved in water to 1 l solution.

[1] Ammediol = 2-amino-2-methylpropane-1,3-diol.

Table 6: pH 8.1–7.7 gel system (5%) for double-disc electrophoresis. Acc. to RACUSEN [R 4].

Gel component		Gel stock solutions				Electrode buffer solution
	Sep. at pH	No.	Components per 1000 ml soln.	pH	Mixing ratio (v/v)	Components per 1000 ml soln.
Cathodic separation gel	7.5	1	1 N KOH 12.0 ml Taurine 7.5 g TEMED 0.25 ml	8.1	2 p. No. 1 2 p. No. 2 4 p. No. 3	Cathode: Taurine 4.75 g 1 N KOH 3.8 ml
		2	Acrylamide 20.0 g Bis 0.8 g $K_3Fe(CN)_6$ 75.0 mg			
		3	Per 0.14 g			
Anodic separation gel	8.3	4	1 N HCl 12.0 ml Imidazole 5.0 g TEMED 0.5 ml	7.7	2 p. No. 4 2 p. No. 2 4 p. No. 3	Anode: Imidazole 2.58 g 1 N HCl 3.8 ml
Spacer and sample gel		5	Safranin O 10 μg Bromphenol-blue 10 μg		4 p. No. 5 3 p. No. 6 3 p. No. 7 3 p. No. 8 35 p. No. 9	
		6	TEMED 1.0 ml			
		7	Riboflavin 4.0 mg			
		8	Acrylamide 40.0 g Bis 10.0 g			
		9	0.25 M phosphate buffer sample material q. s.	7.0		

Table 7: Gel systems for polyacrylamide gel electrophoresis using discontinuous voltage gradients (ALLEN-System) [O 10].

Gel system		Range of application (examples)	Separation gel stock solutions		
No.	Designation		No.	Components per 100 ml solution	pH at + 25° C
1	pH 9-plasma-protein-system	Plasma and serum proteins, esterases	1 (1.5 M)	1 N H_2SO_4 31.0 ml Tris 18.15 g TEMED 0.24 ml	9.0
			2	Acrylamide 32.0 g Bis 0.8 g	
			3	Per 0.21 g	
			3a	Per 0.105 g	
			4 (0.3 M)	Soln. No. 1 20.0 ml TEMED 0.19 ml	9.0
2	pH 9-lipo-protein-system	Plasma lipoproteins	6 (0.75 M)	1 M Citric acid 7.0 ml Tris 9.075 g TEMED 0.24 ml	9.0
			7 (0.15 M)	Soln. No. 6 20.0 ml TEMED 0.19 ml Brij 35, 10% aqueous 0.2 ml	9.0
3	pH 8.5-enzym-system	Spleen extract proteins, esterases, LDH iso-enzymes	9 (1.5 M)	1 N HCl 58.0 ml Tris 18.15 g TEMED 0.24 ml	8.5
			10 (0.3 M)	Soln. No. 9 20.0 ml TEMED 0.19 ml	8.5

Table 7 continued

Mixing ratios (v/v) of stock solutions for a discontinuous gradient gel				**Electrode buffer solution**	
8% sep. gel	6% sep. gel	4.5% sep. gel	8% cap and well form. gel	Components per 1000 ml solution	pH at 25° C
1 p. No. 1 1 p. No. 2 2 p. No. 3a	12 p. No. 1 9 p. No. 2 3 p. H_2O 24 p. No. 3a	8.0 p. No. 1 4.5 p. No. 2 3.5 p. H_2O 16.0 p. No. 3a	1 p. No. 4 1 p. No. 2 2 p. No. 3	Tris 7.86 g Boric acid 1.925 g	9.0
Sample solution (No. 5, 0.075 M): Soln. No. 1 (less TEMED) 5 ml Sucrose 50 g Bromphenolblue, 0.1% aqueous 0.4 ml H_2O to 100.0 ml			pH 9.0		
1 p. No. 6 1 p. No. 2 2 p. No. 3a	12 p. No. 6 9 p. No. 2 3 p. H_2O 24 p. No. 3a	8.0 p. No. 6 4.5 p. No. 2 3.5 p. H_2O 16.0 p. No. 3a	1 p. No. 7 1 p. No. 2 2 p. No. 3	Tris 7.86 g Boric acid 1.925 g	9.0
Sample solution (No. 8, 0.0375 M): Soln. No. 6 (less TEMED) 5 ml Sucrose 50 g Bromphenolblue, 0.1% aqueous 0.4 ml H_2O to 100.0 ml			pH 9.0		
1 p. No. 9 1 p. No. 2 2 p. No. 3a	12 p. No. 9 9 p. No. 2 3 p. H_2O 24 p. No. 3a	8.0 p. No. 9 4.5 p. No. 2 3.5 p. H_2O 16.0 p. No. 3a	1 p. No. 10 1 p. No. 2 2 p. No. 3	Tris 6.0 g Glycine 28.75 g	8.5
Sample solution (No. 11, 0.075 M): Soln. No. 9 (less TEMED) 5 ml Sucrose 50 g Bromphenolblue, 0.1% aqueous 0.4 ml H_2O to 100.0 ml			pH 8.5		

Note to table 7: The samples are dissolved in or diluted with solution No. 5, 8 and 11, respectively. The polarity is: cathode⊖ top, anode⊕ bottom. Staining for gel system No. 1 with a solution of 0.017 g Amidoblack in 100 ml of 10% acetic acid 2 hrs. at 65° C. Staining for gel system No. 2 (lipoproteins) with an alcoholic solution of Lipid Crimson (see page 76).

Table 8: Highly concentrated gel system for the fractionation of oligodeoxynucleotides (20% separation gel) acc. to ELSON and JOVIN [E 4].

Separation gel stock solution			Spacer gel stock solution			Lower	Upper
Mixing ratio (v/v)	Components per 100 ml solution	pH	Mixing ratio (v/v)	Components per 100 ml solution	pH	electrode buffer	electrode buffer
33	Acrylamide 60.0 g Bis 3.0 g		50	Acrylamide 5.0 g Bis 1.5 g		0.1 M Tris-HCl, pH 8.2	0.052 M Tris + 0.052 M glycine, pH 8.9
40	Bis 3.0 g		25	Tris 2.78 g 1 N HCl to pH TEMED 0.1 ml	7.2		
25	Tris 1.815g 1 N HCl to pH Bis 3.0 g	9.0	12.5	Riboflavin 3.0 mg			
2	Riboflavin 3.0 mg Bis 3.0 g		12.5	H_2O			

2.3.2. Preparation of Cylindrical Gels

2.3.2.1. Preparation of the Spacer Gel

Saccharose solution (40 %, 0.2 ml of solution No. 7, Table 4) is filled into a rubber cap and the gel tube is twisted into the cap in a precisely vertical position. The formation of air bubbles is effectively avoided in this manner. Using a 2 ml or 10 ml syringe with a stainless steel needle of 8 cm length or a polyethylene tubing of about 8 cm length (0.6–0.8 mm I.D.), about 0.15 ml spacer gel solution is layered on the sucrose solution. This is covered with about 0.1 ml water. Special care must be taken that the interfaces between the layers are sharp and that no mixing of layers occurs. The resolving capacity of the method is substantially determined by the sharpness of the boundary between spacer and separation gel. Some investigators prefer to apply water with a microcap disposable micropipette (from firm 7) by running the water down the inner side of the tube from a point only 1–2 mm above the gel surface. The spacer gel solution is irradiated with a fluorescent daylight lamp; strong sunlight may be sufficient, too. The distance between lamp and tubes should amount to 2–3 cm. The tubes should not undergo heating during this process. Irradiation proceeds for 20–30 min. Approximately 6 min. after the start of irradiation, an opalescence develops in the gel solution indicating the start of photopolymerization.

2.3.2.2. Preparation of the Separation Gel

Following the photopolymerization of the spacer gel, the supernatant water layer is removed with the syringe and residual water droplets are suctioned off with a wick of lint-free filter paper or celluloseacetate film. Only now will the separation gel solutions be slowly mixed. Strong mixing should be avoided, because oxygen retards the polymerization (p. 3). – In case of retardation or inhibition of gelation, the mixture of separation gel stock solutions should be degassed in vacuo. For this purpose, the mixture is aspirated into a syringe, the syringe opening is closed and an underpressure is produced in the syringe. – The space over the spacer gel is quickly rinsed twice with a small quantity of the separation gel mixture; the rinsing liquid is discarded. The tube is now filled with the mixture until the liquid forms a concave meniscus over the tube rim; no air bubbles may form during pouring, however. (In the case of a long gelation time, the tube edge is covered smoothly with a thin plastic film without air bubbles; a small amount of liquid will run down the outer tube wall.) The time between preparation and pouring of the separation gel mixture should not exceed 5 min. The tube is now allowed to stand for 30–40 min.; gelation normally begins 15–20 min. after casting. After polymerization, the tube is carefully removed from the rubber cap: For this purpose, the rubber mounting is depressed, air is allowed to enter and the tube is slowly tilted out of the cap. The tube is inverted, is loosely set into another rubber cap – the separation gel facing downward – and the saccharose solution is removed. The space over the spacer gel is rinsed with spacer gel solution, the prepared sample solution (p. 68) is applied and the latter is very slowly covered with electrode buffer solution to the upper rim of the tube. Care must be taken that the denser sample solution will not mix with supernatant electrode buffer solution. It is also possible to add this solution first and underlayer it with sample solution; again no sample solution should be lost by entry into the electrode buffer solution. Finally the sample solution is photopolymerized.

If the surface of the spacer gel has a nonuniform (curved) shape, an additional layer of spacer gel of 4 mm height is photopolymerized by the above method under water before the sample is applied. – Nelson and Hale [N 25] select the following sequence of layers (from the bottom of the tube to the top): Separation gel, spacer gel, sample solution, spacer gel. Thus an additional plug of spacer gel on top of the sample prevents upward diffusion and dilution of sample. This technique is particularly recommended if the sample solution should not be photopolymerized for reasons given on page 60. The authors prefer the application of samples in spacer gel solution, since saccharose solution sometimes leads to a nonuniform interface between sample and spacer gel.

2.3.2.3. Alternative Gel Preparation

– In general, the guidelines described above for the preparation of spacer and separation gels should be followed. The reasons for this are given on p.57. However, since a number of laboratories are still using the original instructions of Davis [D 5], this method will be briefly described here: In principle, the separation gel, followed by the spacer gel and finally, a sample gel are polymerized on top of each other.

About 0.2 ml separation gel solution is filled into the rubber cap, the glass tube is inserted, separation gel solution is added to a level of about 4 cm and is covered very slowly with water. After polymerization (about 30 min), the water layer is removed, the space over the separation gel is rinsed several times with spacer gel solution, is filled with the same solution to a height of about 1 cm and again covered with water. After photopolymerization (about 20 min; about 2–3 cm distance from the fluorescent daylight lamp), the water layer is removed, the sample solution (or the sample gel solution) is applied and covered with electrode buffer solution.

2.3.3. Problems of the Preparation of Spacer and Separation Gels

2.3.3.1. Effects of Catalysts and Gel Components on Gel Polymerization, Gelling Time and Electrophoretic Migration

The question may arise why two different catalyst systems should be used for the preparation of the spacer (stacking) and separation gels. According to Davis [D 7], *photocatalyzed gels* have larger pores under the cited time conditions compared to peroxide-catalyzed gels. The catalytic system of riboflavin–TEMED therefore is preferable for large-pore spacer gels. In addition, this system has two further advantages: first, the activity range of riboflavin falls into the 5 ppm range (700 ppm for ammonium persulfate); secondly, the gelation time depends only on the light energy applied and is therefore more easily controllable. However, for highly acid gel systems, riboflavin is not suitable [R 34] (pp. 42, 55). Since the time of photopolymerization affects the gel porosity and hence the mobility of proteins in the gel, this condition should be standardized. According to Brackenridge and Bachelard [B 53], photopolymerization should proceed for ample time (e.g. 60 min.) to achieve completion. In the case of *persulfate – catalyzed* gels the gelling time predominantly depends on the catalyst concentration. For reproducible analyses, these gels should be used within constant periods of time after molding. Apparently aging of the gels exerts an influence on the reproducibility of the separation pattern. Thus Loening [L 36] found that different batches of gels or gels of different ages from the same batch are poor in comparison, while separate gels of same age run at the same time in the same tank furnish identical results. Moreover, Neuhoff [N 27] observed that he could obtain good separations in microgels only if they were at least one day old. Unfortunately, the physicochemical processes of gel polymerization which might explain these phenomena are poorly understood and have to await further investigations.

Kingsbury and Masters [K 19a] have emphasized that the catalyst concentration exerts a notable effect on gelling time and relative mobility of charged particles of both low and high molecular weight. At constant monomer and increasing catalyst concentration the mobility of bromophenolblue decreases, whereas the mobilities of carbonic anhydrase isozymes increase, probably due to displacement of chains. Hence catalyst concentration greatly influences the length of the polymer chains and the number of chains per unit volume, both major factors of gel formation.

It is therefore essential to adopt a standardized set of conditions for electrophoresis, gel components and polymerization.

TOMBS [T 16] pointed out that the polymerization reactions of acrylamide are never completely quantitative, i.e. a certain fraction of unreacted monomer is always present, which can undergo secondary reactions with the sample components to be fractionated. The percentage of free acrylamide depends mainly on the gel concentration. For this reason it may become necessary for special separation problems to *purify* the separation gels from unreacted material (monomer, catalysts) by pre-electrophoresis (removing ions) for hours and/or prewashing (removing ionic and non-ionic compounds) for days. Thus UV-absorbing material can be extracted [L 35, L 36] (see page 41). TABER and SHERMAN [T 1], for example, incubate the separation gels in buffer solution for at least 2 days prior to electrophoresis. The spacer gels are prepared subsequently.

Large-pore stacking or spacer gels are obtained when the polymerization product from the minimum concentrations of monomer and comomer still exhibits mechanical gel properties. Gel solutions with less than 2 % acrylamide and 0.5 % crosslinking agent do not polymerize; but higher concentrations yield reproducible gels. According to DAVIS [D 7], the addition of sugar improves the reproducibility of gel formation and the mechanical gel stability without noticeably modifying the pore size. Spacer gels serve mainly as anticonvection media. RITCHIE et al. [R 44] found that they fulfill two additional functions: They retain sample precipitates (and lipids) from the surface of the smallpore separation gel thus preventing plugging of the pores and reducing band tailing; moreover, they "correct" irregular sample boundaries prior to the actual separation. – Problems arising with the use of photopolymerization and the influence of several additives have been discussed by NEEDLES [N 19a].

If gelation of the separation gel should take significantly longer than half an hour (and the solutions are not overaged), the prescribed concentration of TEMED may be increased by 20–30 % or 0.01 % DMPN (see page 3) is added. These compounds are added to the buffer solution, not to the catalyst solution, because the gel solution otherwise will not gelate. *Delays in polymerization* are unfavorable: generally they indicate impurities in the reagents used. Inversely, too rapid a gelation is retarded by precooling of the solutions or by the addition of potassium ferricyanide. If the polymerization of acidic gel solutions (e.g. gel system No. 9) is retarded or inhibited, the use of silver nitrate (0.1 ml of a 0.6 % solution for 8 ml separation gel mixture) instead of TEMED and a lower ammonium persulfate concentration (from 2.8 % to 0.4 % in solution No. 21) are recommended [N 18]. – For the polymerization at low pH (down to 1.0) a mixture of 0.1 % ascorbic acid, 0.0025 % ferrous sulfate and 0.03 % hydrogen peroxide (from 30 % stock) is recommended as a catalyst system [J 17]. Gel solutions with 0.00045 % riboflavin as the photocatalyst gelate within 10 min. after the addition of 0.018 % sodium sulfite [G 12]. Similarly, by adding sodium sulfite (0.035 %) the amount of ammonium persulfate required for sufficient polymerization can be reduced to 0.04 % in the presence of 0.24 % DMPN [S 66a].

2.3.3.2. Water Layering

The gel solutions are covered with water in order to produce the smoothest possible gel surfaces. Adhesion forces between the gel solution and glass, which lead to the formation of a meniscus, should be suppressed in this manner. As noted earlier (p. 52), this manipulation requires special care, because any distortion in gel surface will be reflected in the pattern of the separated proteins. It will be observed that a small part of the gel solution at the water boundary does not gelate. According to Davis [D 7], this zone of 1–3 mm depth represents a buffer layer which suppresses the diffusion of salts and water between the water layer and the polymerizing gel.

2.3.3.3. High Density Separation Gels

Problems arising with the use of *gels of high acrylamide concentration* (20 %) have been discussed by Elson and Jovin [E 4].

Main difficulties concern inhomogeneous polymerization and separation of the gel from the mold walls. More homogeneous gels are obtained by reducing the rate of polymerization (omission of TEMED, photopolymerization with riboflavin instead of chemical polymerization with persulfate). Swelling of the lower gel surface is prevented by increasing the Bis concentration from 0.4 to 3 %. To produce a flat upper gel surface isobutanol instead of water is layered over the separation gel solution. Isobutanol should be immediately removed after polymerization in order to avoid dehydration of the gel. Finally, highly concentrated gels produce more ohmic heating than e.g. 7 % gels, thus lower current densities (2 mA/cm^2 and less) must be applied. See also [P 15].

2.3.3.4. Sources of Error in Faulty Separations

Aside from the fact that good separations are possible only in gels consisting of high-purity components, the following influences can lead to unsatisfactory results [R 43–44]: "Aging" of the gel solution, improper pore size (gel concentration), unsuitable buffer system (buffer ions, pH range, ionic strength), too short a migration distance (separation gel), insufficient concentration of the sample components prior to separation (too weak a pH discontinuity, too short stacking gel layer), careless application of water layers, nonuniform gel polymerization (heterogeneous pores, air bubbles in gel), alterations during storage; excessive ionic strength of the sample solution (page 68), mixing of sample solution with supernatant buffer, protein-protein interaction due to excessive protein stacking, irreversible aggregation, sample precipitation (especially at the boundary of spacer and separation gels) which dissolves only slowly (tailing), blocking of the boundary by high molecular weight substances, insufficient adhesion of the gel to the mold and thus, penetration of sample solution into the "leak" (tailing) (see page 39); diffusion and ohmic heating; catalyst artifacts (page 59), reactions of the molecules to be separated with gel components (monomer, catalysts).

2.3.4. Modifications of Gel Compositions and Suggestions for Special Separation Problems (Modifications of Disc Electrophoresis)

2.3.4.1. Sample, Spacer and Separation Gel: Sequence of Preparation

According to the original instructions of ORNSTEIN and DAVIS [D 5], the following gels are polymerized on top of each other: first the separation gel, then the stacking gel and finally, a sample gel. This sequence differs from that described on pp. 52-53. DAVIS [D 7] explains this as follow.

Although a meniscus is avoided by a layer of water, distorted gel zones can form in the separation gel, particularly at its boundary with the spacer gel, which may cause curved protein bands. During gelation, the volume of the gel solution decreases; furthermore, the polymerization may have an irregular course. The newly described method prevents this phenomenon to a large extent. Distorted bands can also be caused by too rapid removal of the rubber cap (if it covers the separation gel), by overaged monomer and comomer solutions as well as by too rapid gel polymerization. In the latter case, the concentration of TEMED may be reduced.

2.3.4.2. Band Curvature and Temperature Effects

Clearly, microdensitometer scanning requires a minimum of *band curvature* to record close bands which are visually distinct. Band curvature may be caused by two effects. The more common and frequent effect arises from the *temperature gradient* set up during electrophoresis from the central axis to the periphery of the gel [P 3]. Enhancement of ohmic heating in the central axis inevitably leads to an increase of protein mobility. Excess ohmic heating may be prevented by reducing the ionic strength, the gel dimensions (particularly the gel thickness) or the current density. In practice only the latter is varied. The permissible level of ohmic heating consistent with a reasonable amperage is mostly determined empirically, but may be calculated (see page 21). Cooling of any electrophoretic medium is only efficient, if the generated heat can be dissipated rapidly enough.

KNUDSEN et al. [K 23a] have measured the temperature rises (from 17° C) in the different gel parts of the standard gel system No. 1a (pp. 44-45) during 60 min. of electrophoresis at 4 mA/gel: sample gel 32°, stacking gel 44°, upper part of separating gel 38°, lower part 23° C. Cooling reduces the temperatures, but the stacking gel may still attain 32° C. Interestingly however, the thermolabile LDH 5 isoenzyme was not inactivated by the temperature rise.

The second cause of band curvature may be a *too rapid polymerization* leading to nonuniform contraction of the gel with pronounced adhesion to the glass wall. It may be prevented by coating the inside walls of glass tubes with a film of silicone prior to filling with gel solutions [S 40]. The tubes are dipped into "General Electric SC–87–Dri-Film" or into a 1 % solution of dimethyldichlorosilane in benzene and heated to 60° C [S 29]. However this method is not generally recommended since rearband flatness in complex mixtures was not found to be reliably reproducible. Furthermore leakage of the sample due to insufficient adhesion of the gel to the silicon wall may occur.

2.3.4.3. The Use of the Sample Gel

According to Davis [D 7], incorporation of the sample into a gel prevents thermal convective currents and yields somewhat higher resolution. It assures an orderly movement of proteins into a stack without backconvection into the upper buffer. This method has been tested against others, e.g. layering the sample directly under buffer onto stacking gel, with added sucrose and/or monomer, and found to be most reliable and least technique-sensitive. However, in some cases [H 16, H 27, H 33, N 15, P 24] gelation was prevented. It was argued that irradiation and polymerization (formation of free radicals) may lead to uncontrolled reactions such as denaturation. If this is a matter of concern, the incorporation of the sample into a solution undergoing vinylpolymerization should be avoided.

In any case, care must be taken that the sample solution will not mix with supernatant electrode buffer and be diluted or lost as a result. For this reason, the density of the sample solution is increased by the addition of saccharose (p. 68). Stock solution of basic proteins should not be prepared with saccharose, however, since the ε-amino-group of lysine can react with carbohydrates [H 6, G 25]. For other properties of the sample solution required for optimal separation conditions see page 68.

2.3.4.4. Modification of the Spacer Gel

According to Broome [B 58], a suspension of Sephadex G–200 in 20 % saccharose and spacer gel buffer (*Sephadex concentration zone*) can also be used in place of the spacer gel. This method is particularly recommended, if artifacts by free unreacted monomer molecules, riboflavin or persulfate are suspected (see chapter 2.3.4.6.). The sample is mixed with the suspension and applied on the Sephadex stacking layer. This reduces the working time (gel preparation and electrophoresis) considerably. Instead of Sephadex, agarose in 2 % solution is also suited for the preparation of a spacer and/or sample gel

[M 48]. The proteins do not necessarily have to be applied in solution, since *paper strips* impregnated with protein solution and dried subsequently can be imbedded in the Sephadex layer (cf. p. 64). In the subsequent electrophoresis, the proteins are eluted through the migrating boundary and are separated. According to MATOLTSY [M 23], a similar method even permits the extraction and separation of proteins from small quantities of macerated tissue (horny material).

In place of the homogeneous layer of sample solution or gel, two glass bead beds of 2 mm thickness (about 0.2 mm bead diameter) are prepared between which the sample powder, suspended in spacer gel solution, is imbedded. The space over the upper glass bead bed is filled with spacer gel solution up to the rim of the tube. Finally, photopolymerization is allowed to take place as usual.

2.3.4.5. Suspension of the Spacer Gel and Pre-Electrophoresis

In some cases, the spacer gel was dispensed with [e.g. B 6, C 34, C 41, F 16, H 27, M 34, V 3]. Consequently the concentration effect on the basis of a discontinuous pH and voltage gradient was abandoned. Hence these techniques are not disc electrophoretic. However, for the reasons discussed below, they are mentioned.

FESSLER and BAILEY [F 16] found a 2 hr. *pre-electrophoresis* of the separation gel, prior to sample application, necessary to avoid artifacts which occur during the migration of collagen components. VEIS and ANESEY [V 3] observed similar interferences. PANYIM and CHALKLEY [P 3] obtained poor resolution and reproducibility of histone patterns when pre-electrophoresis had been omitted. MAIZEL [M 2] found that preparative gel electrophoresis can produce higher yields if the small-pore gel is first subjected to pre-electrophoresis for a few hours in separation gel buffer. Similarly HJERTÉN et al. [H 26] pre-electrophoresed to remove ammoniumpersulfate and UV-absorbing impurities. The efficiency of pre-electrophoresis in removing persulfate can be tested by incubating the gels in benzidine reagent (2% in 10% acetic acid) [B 18].

Evidently, uncontrollable reactions with sample materials can be suppressed, if ionic impurities of the gel components and unreacted molecules of the polymerization process are removed prior to sample application by *pre-electrophoresis*. In addition pre-electrophoresis sets up a constant electric resistance inside the gel. In acid systems e.g., the initial electrophoresis causes, at constant current, a drop in pH and ionic strength and consequently an increase of voltage and resistance which stabilize with time [P 3].

2.3.4.6. Persulfate and Irradiation Artifacts

BREWER [B 56] observed that *ammoniumpersulfate*, in the presence of 8 M urea, apparently inactivates yeast enolase with the formation of a protein cleavage product. Tyrosine and tryptophan residues as well as other oxidizable residues, such as sulfhydryl groups (even without 8 M urea) are attacked. Similar persulfate-caused oxidation products were found by MITCHELL [M 53] in clostridial peptidase B, while FANTES and FURMINGER [F 7] observed a loss of biological activity of interferon poliovirus and ribonuclease under the influence of persulfate. The effect was more pronounced in low pH (4.3) than in high pH (8.9) runs and could be avoided by using riboflavin as catalyst.

ORNSTEIN [O 9a] suggested that the mechanism of this phenomenon may be an indirect result of, rather than a direct attack by, the catalyst. Since persulfate has a far greater mobility in the gel than any protein, it migrates ahead and never comes in contact with the protein. It does, however, eliminate all reducing agents and then leaves the gel in an oxygen-rich state.

To avoid such *persulfate artifacts* in disc electrophoresis, BREWER [B 56] recommended the use of riboflavin-photopolymerized gels or the addition of thioglycolate – a reducing agent which migrates with the buffer boundary – to persulfate-catalyzed gels. ORNSTEIN [O 9a] also advised the application of an anionic reducing agent ahead of the sample or where protein requires continuous input of SH-rich environment, the addition of 1/100 to 1/1000 part of thioglycolate per part of glycine to the upper buffer. However, thioglycolate is only suited for proteins (e.g. cysteine-free) which do not react with it. Moreover, KOHN [K 28] cautioned that thioglycolate may damage certain enzymes, because of its acidic nature, as shown by the color change in the tracking dye. Alternatively KOHN proposed the incorporation of 5 mM mercaptoethanol into the spacer gel and sample or in the upper buffer.

On the other hand photopolymerization of the sample gel containing sensitive proteins may lead to *irradiation artifacts* due to the formation of molecular aggregates or to disaggregation or chain fragmentation. Thus PASTEWKA et al. [P 5b] found an extensive alteration in the electrophoretic pattern of different hemoglobins after exposure to fluorescent light. Finally, NEEDLES [N 19a] pointed out that the formation of hydrogen peroxide during photoinitiation and polymer photografting may introduce artifacts detectable during protein separations.

Poor resolution and inadequate reproducibility may also preclude photopolymerization as a routine substitute for persulfate polymerization. King [K 17a] therefore recommends the addition of antioxidants (dithiothreitol or mercaptoethanol) to persulfate polymerized gels.

2.3.4.7. Solubilizable Polyacrylamide Gels

Solubilizable polyacrylamide gels can be obtained if the cross-linker N,N'-methylenebisacrylamide (Bis) is replaced by *ethylenediacrylate* (*EDIA*) or *N,N'-diallyltartardiamide* (*DATD*). EDIA-gels [C 34] are solubilized at alkaline pH (see page 97), whereas DATD-gels dissolve in 2% periodic acid in 30 min. at room temperature [A 29a]. The properties of DATD-gels are very similar to Bis-gels. DATD is easily prepared from diethyltartrate and allylamine, and may be obtained e.g. from firm 30.

2.3.4.8. Gradient Gel Electrophoresis

According to SLATER [S 45], certain plasma and cell proteins can be better resolved in *polyacrylamide pore gradients*. The method deserves attention, since it offers several possibilities to utilize the potentials of polyacrylamide gel electrophoresis: Different acrylamide and/or N,N'-methylene-bis-acrylamide concentration gradients can be formulated; moreover, the form (linear, stepped, exponential, etc.) and slope of the gradients can be modified. SLATER [S 46] pointed out, that the migration rate of proteins in a linear gradient gel decreases with time; eventually a particular protein will reach "its" pore limit. By migration to the position set by the pore size of the gel a band of increased sharpness is obtained. The band pattern remains stable although the electrical field is maintained. *Pore limit electrophoresis* thus provides a valuable means for separating by size fractionation a protein mixture of widely differing sizes.

Other investigators [E 5, M 11a–13, P 27, T 10] have also emphasized the significance of *continuous gradient gels*. Moreover MARGOLIS and KENRICK [M 11–13] designed units in which reproducible continuous gradient gels (containing 4–24 % acrylamide) may be prepared.

Linear concentration gradients are produced by means of two connected vessels filled with calculated quantities of gel mixtures (Table 8). Ammonium persulfate solution is added immediately before pouring into the vessels. The layering device consists of a mold of rectangular cross-section with the lower segment forming a funnel tapering down to the inlet opening. Tubes or assembled flat cells are stacked vertically on a supporting net platform. The device is first fed with water to cover the cells, then with gel gradient mixture and finally with 25 % sucrose solution to displace the gel mixture from the lower part of the layering device into the gel molds. To avoid convection currents due to the exothermic polymerization reaction, gelation must start at the top and progress downward. This is achieved by establishing a concentration gradient of catalysts along the column. – Moreover MARGOLIS [M 11] constructed a versatile device for generating continuous concentration gradients with variable profile and volume. A similar apparatus with fixed profile was described by PRATT and DANGERFIELD [P 27] for the analysis of lipoproteins.

Table 9: Composition of solutions for continuous gel concentration gradients with a continuous buffer according to MARGOLIS and KENRICK [M 12].

Reagent	Solution A (26 %)	Solution B (4 %)
1. Acrylamide	76.57 g	11.78 g
2. N,N'-Methylenebisacrylamide	4.03 g	0.62 g
3. Sucrose	12.4 g	3.1 g
4. Buffer, pH 8.28 (see below)	to 302 ml	to 302 ml
5. Dimethylaminoproprionitrile (DMPN)	0.03 ml	0.09 ml
6. Ammoniumpersulfate, 10 % aqueous	7.75 ml	7.75 ml
Total volume	310 ml	310 ml

Buffer: Tris 10.75 g, EDTA-Na_2 0.93 g, boric acid 5.04 g, H_2O to 1 l. The reagents 1–6 are dissolved with gentle heating, degassed under suction and cooled to 18° C.

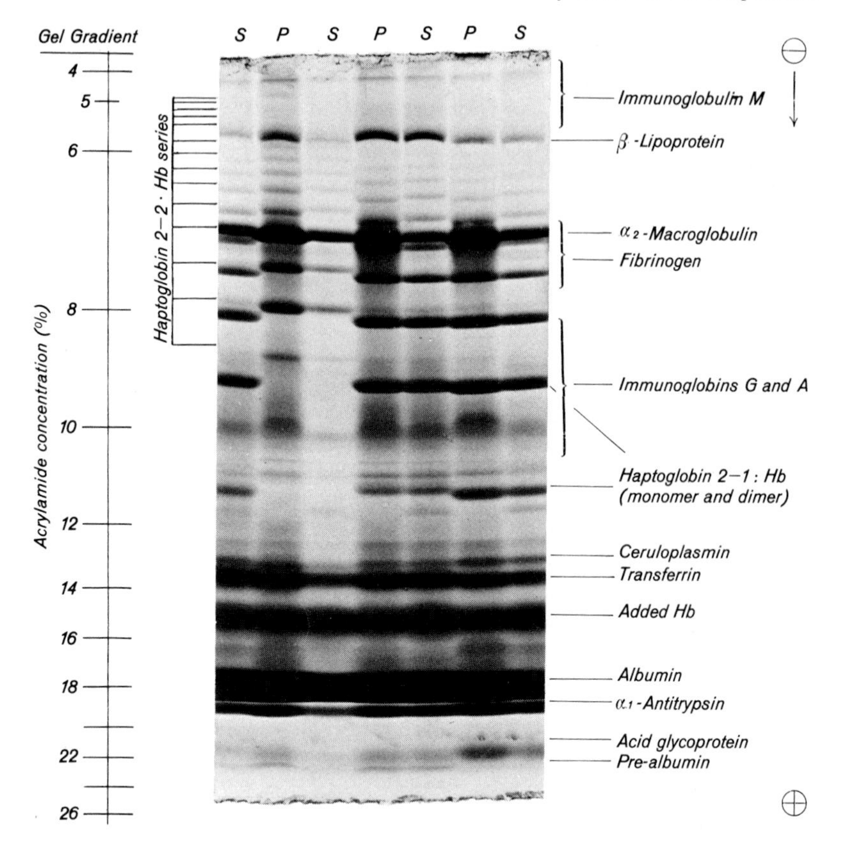

Fig. 21. Electrophoresis of human plasma (P) and serum (S) proteins in a 4–26 % concave gel gradient acc. to MARGOLIS and KENRICK [M 12].

Tris – borate – EDTA buffer pH 8.2 acc. to KITCHIN [K 22], sample amount approx. 300 μg total protein contained in approx. 5 μl, electrophoresis at 70 V, 10 mA (final) and 15° C for 24 hrs, Amido Black staining.

For continuous 4–24 % gradient gels, suitable to fractionate serum and plasma proteins (Fig. 21), Margolis and Kenrick used the solutions listed in Table 9.

Using a simple gradient mixer made from syringes, needles, tubing and a rubber cap (Fig. 22), KIDBY [K 15a] prepared linear gradient gels in small columns. The fixed volume of the mixer required an approximation of exponential to linear gradients for which the author elegantly accounted.

A simpler, yet efficient method to produce gradients gels consists of putting two or more separation gels on top of each other with decreasing acrylamide concentration in the upward direction (*discontinuous gradient gel electrophoresis*). Such gel combinations proved to be useful for the fractionation of e.g. serum protein, [W 36–37], collagen components [C 40] and RNAs [G 43] (Fig. 80). Utilizing the apparent sharpening of boundaries as they pass gel concentration discontinuities, the resolution of this method can be highly improved so

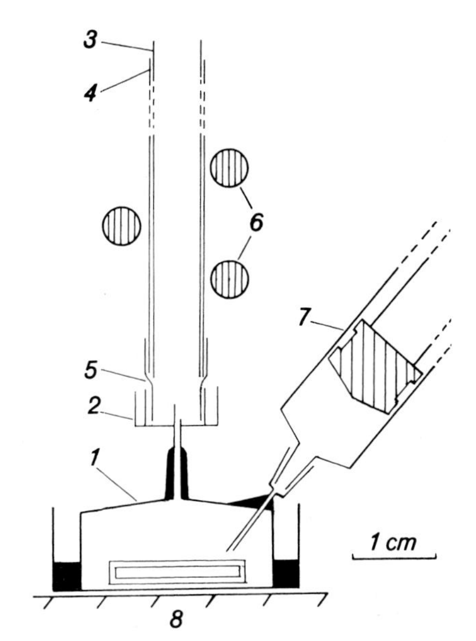

Fig. 22. Apparatus for the preparation of gel gradients in small gel columns.

(1) Mixing chamber made from a 35 ml disposable syringe with outflow needle cemented into syringe end; (2) rubber serum bottle cap; (3) thick-walled Tygon tubing; (4) glass tubing; (5) thin-walled Tygon tubing; (6) clamp; (7) syringe containing monomer solution with needle; (8) magnetic stirrer. From: KIDBY [K 15a].

that differences between both high and low molecular weight components can be detected [C 40].

However gradient gels first should be used with already well-charted and familiar protein mixtures to avoid misinterpretation. In addition, gradient gels – particularly continuous – should be checked for slope homogeneity and reproducibility.

2.3.4.9. Two-Dimensional Polyacrylamide Gel Electrophoresis

Proteins of similar charge and size may be superimposing in a gel electropherogram, particularly when complex mixtures of numerous proteins are to be analyzed. Better resolutions may be obtained, if the electropherogram is subjected to a second separation: After the first run the gel is removed from its mold and horizontally embedded in a gel slab, thus serving as starting gel for the second dimension. Slab gel apparatus can be used for this purpose. RAYMOND [R 11] called his method Orthacryl technique. KALTSCHMIDT and WITTMANN [K 3a] showed that the two-dimensional system may serve as an excellent fingerprinting technique, e.g. for the about 50 ribosomal proteins from Escherichia coli. Preferably the second run should utilize another separation principle, for example separation according to size or isoelectric point only. Thus, gel electrophoresis has been combined with gradient gel electrophoresis [M 13] and gel isoelectrofocusing [D 2a, D 27a, M 1, W 35] (see page 132).

The refined two-dimensional gel slab technique by WEIN [W 5a] produces sigmoid gradients with respect to the migration direction of the samples. Such gradients are obtained by filling a rectangular gel compartment in the diagonal direction with a solution of exponentially increasing acrylamide concentration. In practice, the compartment is tilted at 45° angle and filled through a port in the bottom corner. Both runs are performed in the same gel, thus avoiding any additional manipulation.

2.3.4.10. Two-Directional (Double) Disc Electrophoresis

According to Clarke [C 41], *retrograde ascending* components, migrating with the opposite charge at the respective pH value, can be detected in an additional separation gel of about 2 cm length above the sample solution (or sample gel). Racusen [R 4] has expanded this technique into the so-called *double-disc electrophoresis*: in this procedure, the cationic and anionic components of the same material are concentrated and separated simultaneously. The substance is concentrated in the sample and spacer gel which is located between a cathodic and an anodic separation gel (Fig. 23). Taurine ($pK_2 = 8.7$) and its potassium salt is the buffer for the cathodic gel and imidazole ($pK = 7.0$) and its hydrochloride for the anodic. The composition of the double-gel system is listed in Table 6 (p. 49). A similar technique was used by Felberg and Schultz [F 11] for the separation of myeloperoxidase proteins.

Two tubes, each containing a separation gel are connected by plastic tubing and the sample-spacer gel solution is injected (about 1 ml) into the space between the two separation gels. Care must be taken to prevent entrainment of air bubbles; these are allowed to escape through an injection needle. It is also possible to pierce the tubing and to charge the solution with a pipette. The perforation is sealed with a self-adhesive plastic film. Finally photopolymerization is carried out as usual.

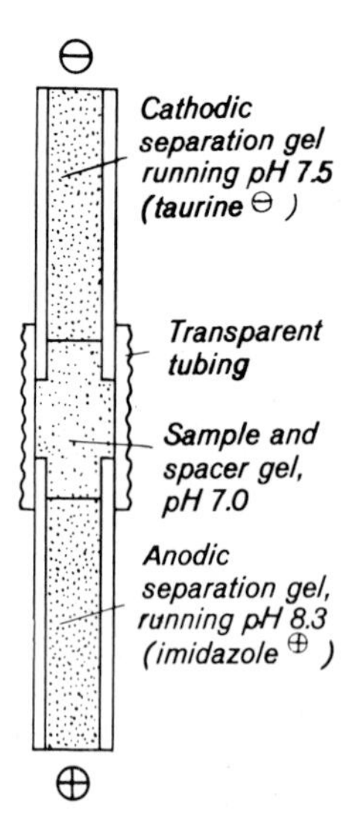

Fig. 23. – Two-directional (double) disc electrophoresis according to Racusen [R 4].

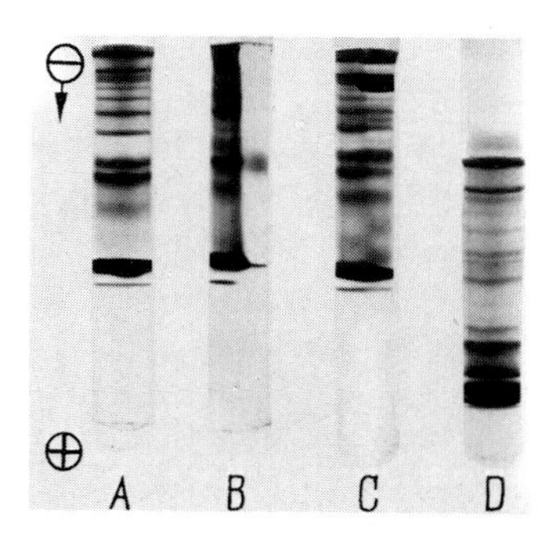

Fig. 24. – Simultaneous electrophoresis of two samples in one gel tube for an exact identification of separated fractions (*Split gel technique*). From: CLARKE [C 41].

(A) Normal human serum. (B) Right: hemoglobin, left: normal human serum. (C) Right: nephrotic serum (with pronounced α_2-macroglobulin fraction at 12 α_2; left: normal human serum. (D) Normal human serum with upper gel (2 cm) for the detection of proteins migrating in opposite directions (ascending retrograde technique, p. 64). – 5 % separation gel with Tris-glycine buffer, pH 8.1 (no spacer gel). Amido Black staining.

2.3.4.11. Sephadex Sandwich Disc Electrophoresis

According to KOPPIKAR et al. [K 31b] a combination of polyacrylamide and Sephadex gel may result in better electrophoretic separations of certain labile macromolecules such as lipoproteins. In this technique, a Sephadex G-200 layer is sandwiched between two separation gel layers, and the lipoproteins are allowed to migrate into the lower separation gel.

2.3.4.12. Split Gel Technique

For accurate comparison purposes, two sample solutions may be analyzed simultaneously in a cylindrical tube by the *double run or split gel technique* (Fig. 24).

According to CLARKE [C 41], a fitted rectangular piece of wax paper (5 × 15 mm) is pressed into the spacer and separation gel to a depth of about 1 mm and both chambers are filled with sample solutions. If the separated bands overlap, the thin, water-impermeable separator is inserted down to the lower end of the tube before gelation and the gel systems are prepared subsequently in the two tube halves. A similar method was developed by LEBOY et al. [L 12] and proved to be particularly suited in comparative studies of the numerous ribosomal proteins of E. coli (p. 173). According to MILKMAN and GERSHWIN [M 51], a water-impermeable separator is needed if the sample solutions are adsorbed on filter paper and the impregnated pieces of paper are inserted into the space which is normally filled by the sample solution or the sample gel. If denaturation need not be feared, the paper pieces may also be dried. The space between is filled with spacer solution and photopolymerization is carried out. JOHNS [J 8] applies this sample solutions to opposite sites of one paper disk. To prevent diffusion into contact a line of Silicone Repelcote water-repellant (available from firm 16) is drawn across the diameter (7 mm) of the disc using a fine glass capillary.

2.3.4.13. Disc Re-Electrophoresis

Fractions which were poorly separated by a single disc electrophoretic run can be resolved further by *disc re-electrophoresis* in gels of different pore sizes according to ROTHMAN and LIDÉN [R 57] (Fig. 25).

The sample, suspended in Sephadex G-200, is applied and run in a 7.5 % separation gel. The fraction is cut from the gel, homogenized and, after electrophoresis in a new 7.5 % gel, collected at the lower end of the tube: for this purpose the rim of the tube is covered with a dialysis membrane in such a manner that a space of about 0.3 ml volume is formed between gel and membrane. The collected fraction finally is suspended in Sephadex G–25, centrifuged and the supernatant, after addition of Sephadex G-200, subjected to re-electrophoresis in a 15 % separation gel.

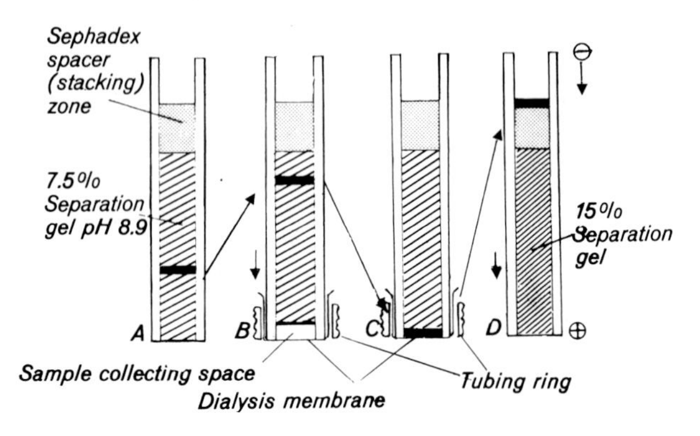

Fig. 25. – Disc re-electrophoresis according to ROTHMAN and LIDÉN [R 57]. (A–D) Sequence of electrophoreses.

2.3.4.14. PH-Reversal Method

JOLLEY and ALLEN [J 14] developed the *pH-reversal method* by which protein complexes (e.g. of basic proteins and globulins) can be analyzed. The run is first carried out at pH 8.6 (modified gel system No. 2) whereby the complex remains stable and is freed from free basic fractions. The complex is then split at pH 3.8 (gel system No. 7) causing the basic components to migrate to the cathode.

Similarly, the preparative method by HILL and WAKE [H 20b] utilizes a pH change, by changing the buffers, to resolve two proteins differing by only one charge unit. Initially, both species are optimally separated by electrophoresis at a pH in close proximity to the isoelectric point of one component. The separated species are then sequentially accelerated off the gel by electrophoresing in a new buffer of different pH and with a mobility in the opposite sense to that of the proteins.

2.3.4.15. Agarose-Polyacrylamide Composite Gels

The *mechanical properties* of the polyacrylamide gel (such as elasticity, strength, elongation and viscosity) are reported to be clearly improved by the addition of 0.8 % *agarose* according to URIEL [U 1]. The advantages become particularly evident when gels with acrylamide concentrations of less than 3 % must be used. The polyacrylamide gel may be prepolymerized first at 50° C for 1–5 min; the temperature is then lowered and the agarose allowed to gelate in the polymer. Suitable concentrations: 3–9 % acrylamide, 0.08–0.24 % Bis, 0.01–0.06 % ammonium persulfate and 0.06 % TEMED. Even < 2 % acrylamide containing gels can be handled without difficulty if they contain a 0.5 % agarose "corset". These composite gels are, according to PEACOCK and DINGMAN [P 9], excellent sieving matrices for high molecular weight substances such as nuclear RNAs (see p. 182).

2.3.4.16. Gel Electrophoresis with Detergents

Dissociating or liberating agents in the gel buffers may be useful to release components from protein complexes and to eliminate aggregation. VITO and SANTOMÉ [V 6] reported that association and dissociation of proteins can be studied with the use of 0.5 % *sodium dodecyl sulfate* (SDS, sodium lauryl sulfate) if an additional spacer gel is polymerized over the sample gel as an reservoir; thus the detergent concentration in the gel remains constant during electrophoresis. For staining, see page 74.

SALTON and SCHMITT [S 1–2] demonstrated that proteins can be gently liberated from lipids, lipid soluble constituents (carotenoids, sterols, fatty acids etc.) and cell membranes by inclusion of *deoxycholate* in the sample, which frees the lipid to run in a well-defined band ahead of the protein. Thus a method is indicated for the possible production of functional entities which do not survive defatting with organic solvents. As shown by LEWIS et al. [L 26a] for extracts from pituitary glands, liver and muscle, protein-dissociating fatty acids may be already present in tissue extracts and alter the electrophoretic mobility of proteins. Therefore caution should be used in interpreting such disc patterns. COUTINHO et al. [C 52] reported on the effects of *Triton X-100* on electrophoresis of alkaline phosphatases and esterases. In addition, *urea* and stronger agents like *phenol-acetic acid-water-mixtures* have been used to solubilize and dissociate firmly bound proteins (see page 14). However these are continuous buffer systems and therefore do not require spacer gels.

LIM and TADDAYYON [L 31a] dissolved brain membrane proteins in a solution of 8 M urea, 10% mercaptoethanol, 5% Triton X-100 and 50 mM K_2CO_3 and electrophoresed in gels containing 5 M urea and 0.25% Triton X-100 with a discontinuous acidic buffer system.

2.3.4.17. Disc Electrophoresis of Fluorescent-labeled Proteins

A heterogeneous mixture of *fluorescent – labeled proteins* (fluorescein isothiocyanate and dansylchloride) can be well separated by preparative disc electrophoresis into fractions according to the degree of labeling [K 16]. Separation takes place on the basis that the introduction of fluorescein or dansyl onto a protein alters the net charge by – 3 and – 1, respectively. Thus, more heavily labeled species migrate more rapidly in the gels.

2.3.4.18. Disc Tissue Electrophoresis

To study the proteins and enzymes of small tissue samples MITTELBACH et al. [M 54–55] subject frozen tissue sections of biopsy material (mammalian skeletal muscle) directly to disc electrophoresis. Two sections (25 mm^2 area, each 40 μm thick) are embedded in stacking gel buffer between two layers of stacking gel. The technique, named *disc tissue electrophoresis*, is quick, convenient, highly reproducible and avoids cumbersome steps of extract preparation such as homogenisation and centrifugation.

2.3.4.19. Detection of Biological Activity of Electrophoresed Proteins by Tissue Culture on Gels

For the *detection of biological activity*, HOFFMANN et al. [H 31–32] incubate disc gels containing fractionated active substances with tissue culture explants and observe the influence of the separated components on the cells growing directly on the gels. The method is particularly interesting for cell and tissue culture studies.

In principle, following electrophoresis, the gel is cut into longitudinal strips (p. 100), one of the gel strips (0.5 mm thick) is soaked for 30 min. in Hanks solution containing 0.01 % Terramycin, drained and surrounded along its length with tissue explants in a teflon molding waxed to a glass slide. It is covered with culture medium, incubated in the closed chamber at 37–38° C for 2 days, and observed in a phase contrast microscope. Another gel strip is stained with Amido Black to locate proteins. – The gel may also be cut transversely and the individual discs tested for biological activity.

2.3.5. Preparation of Sample Solution

If possible, the sample solution should have the same buffer, pH value and ionic strength as the spacer gel solution. Consequently, *solid substances* are dissolved in spacer gel solution. For a standard gel cylinder (diameter 5 mm), a concentration of about 1 mg/ml is selected. With a small number of bands (up to about 6), 5–50 μg protein can be readily detected per gel. However, 0.4 μg purified serum albumin may still be detected with the naked eye after staining with Amido Black. The optimum quantity of sample to be applied therefore must be determined experimentally for each case. In most cases, it will range between 10 and 100 μg. To inhibit mixing of the sample solution with supernatant electrode buffer, the density of the sample solution is increased by the addition of sufficient saccharose to form an approximately 1 molar solution (34 g/100 ml). Glycerol, Sephadex G-200 or, for some specific purposes, urea (p. 67) may be used instead of saccharose.

If the ionic strength of the sample solution is too high, blurred boundaries or no separations at all (*salt effect*) are obtained; high salt concentrations furthermore inhibit the polymerization of the sample gel. It is therefore best to desalt prior to electrophoresis. In any case the conductivity should be lower than that of the running gel to provide a sufficient voltage gradient for zone sharpening. Moreover, the medium constituents should not inhibit the free migration of the molecules to be separated, for example by adsorption. In addition, for precise comparison of samples it is essential to equalize sample volumes and ensure that the final sucrose concentration exceeds 5 % [B 53].

Biological fluids are diluted with spacer gel solution such that the desired component is contained in about 0.15 ml total solution. In the case of blood sera, 3 μl (about 200 μg protein) are mixed with 0.15 ml spacer gel solution. With fluids of *low sample concentration*, the spacer gel solution is prepared in a suitable higher concentration by reducing the quantity of water; for example, the spacer gel solutions are mixed with the fluid as follows (Table 4, gel sytem No. 1): 1 part each of solutions No. 4, No. 5, No. 6, 4 parts No. 7, and 1 part liquid; with higher dilution, solution No. 7 is replaced by 2 parts of a 65 % saccharose solution and 3 parts liquid. Of *very dilute fluids*, such an amount is used that the sample solution contains 10–50 μg of material. This may increase the volume of sample solution. In this case, the volume of spacer gel must also be increased, thus requiring the use of longer gel tubes.

2.3.6. Measuring of the Sample Solution

For reproducible, quantitative analysis, it is important to measure the sample solutions accurately. According to St. Groth et al. [S 74], a serious source of error in zone electrophoresis procedures may reside in the precision of measuring microliter quantities of the sample solutions to be applied. With some practice, it is possible to release the microvolumes with a relative standard deviation of ± 5 %. It is immaterial whether microliter (lambda) pipettes or microliter syringes are used. Polyethylene micropipettes are of advantage because they are unwettable and unbreakable. Manufacturers of micropipettes: For example the firms 2, 7, 9; of microsyringes: Firm 14.

Narayan et al. [N 17] carried out systematic statistical studies on methods of sample application, and obtained best reproducibility when an additional spacer gel as a plug was photopolymerized over the concentrated sample solution.

2.3.7. Electrophoretic Procedure

Although separation (small-pore) gels can be stored under buffer at 5° C for a week without loss of resolving power, sample and stacking gels should be subjected to electrophoresis within an hour after their preparation. It is important to remove the gel tubes slowly and carefully from the rubber caps. The gels should not be pulled or twisted; nor should they strip from the inside wall of the tube. The rubber caps are depressed, air is allowed to enter and the tubes are slowly tilted out. Subsequently, they are slowly turned upward into the rubber-grommet lined bores of the upper electrode reservoir until the upper tube rim projects about 5 mm above the grommets. Precooled (+ 4° C) electrode buffer solution is slowly poured into the upper and lower electrode reservoir and special care is taken to prevent the formation of air bubbles at the tube rims. If necessary, air bubbles are removed with the pipette. (In these manipulations, samples which have not been polymerized in a sample gel may be lost into the upper electrode buffer solution due to vibration and turbulence!) The upper electrode reservoir is slowly lowered until the tubes are immersed to a depth of about 4 cm in the buffer solution of the lower rerservoir. Again, no loss of contact due to air bubbles should occur. It may be advisable to "hang" drops onto the lower tube edges before they are immersed in the lower electrode buffer.

In the case of *alkaline* buffer systems (e.g. gel system No. 1), about 0.5 ml of a filtered 0.001 % aqueous bromophenol blue solution, is added to 250 ml of the upper buffer solution, while *acid* systems receive an addition of about 0.5 ml of a 0.005 % aqueous methyl green, methylene blue or Pyronine-Y solution. The dye serves as a marker of the migrating boundary during electrophoresis (pp. 27).

The electrodes are now connected with the dc power supply and the current is adjusted to 1 mA for the first 2 min, and then to 2–5 mA/gel for gels of 5 mm diameter. Generally, currents higher than 5mA/gel should be avoided because otherwise too much ohmic heating is generated. Under normal conditions (e.g. 4 mA/gel, 20° C, about 40 min.), the separation gel attains 35–40° C. The temperature rise can be prevented by reducing the current intensity and extending the running time; for example, with a current intensity of 1 mA/gel, about 150 min. are needed. It should be recalled, however, that high voltage gradients and short running periods generally yield better resolution. Electrophoresis may be stopped when the migrating schlieren boundary (interface), indicated by the dye marker, has moved a distance of about $^2/_3$ to $^3/_4$ of the gel length, depending on the gel system used and the nature of sample material.

Following electrophoresis, the two electrode buffer solutions are decanted. They may be reused in subsequent electrophoresis runs. However, they should not be mixed; nor may the anode solution serve as the cathode solution in the following electrophoresis; thus, in the case of gel system No. 1, for example, the chloride and catalyst ions which have entered the solution would change the ionic composition of the cathode solution.

2.3.8. Detection of Separated Substances by Staining

2.3.8.1. Removal of Gels from Gel Tubes

The separated components are visualized by incubating the gels with fixative-stain solution or substrate solution (for example, for enzyme detection). To prevent a broadening of the sample zones in the gel by diffusion, incubation proceeds immediately after electrophoresis. For this purpose, the gels must be removed from the tubes. This is technically one of the most difficult manipulations of disc electrophoresis; however, with some practice it is effortless even with hard gels. It requires a 2 ml or 10 ml syringe with attached stainless steel needle (about 8 cm long), from which the tip has been clipped and the end filed smooth, plus a little patience. The syringe is filled with water, 50 % aqueous glycerol or ethylene glycol. While continuously rotating the gel tube with the left hand, the injection needle is slowly pushed forward between the inside wall and gel past the spacer and separation gel with the right hand (Fig. 26). The discharging content of the syringe serves as a lubricant. The needle is thus moved spirally along the inside wall of the tube. The gel is loosened and is finally pushed out of the tube by light side pressure

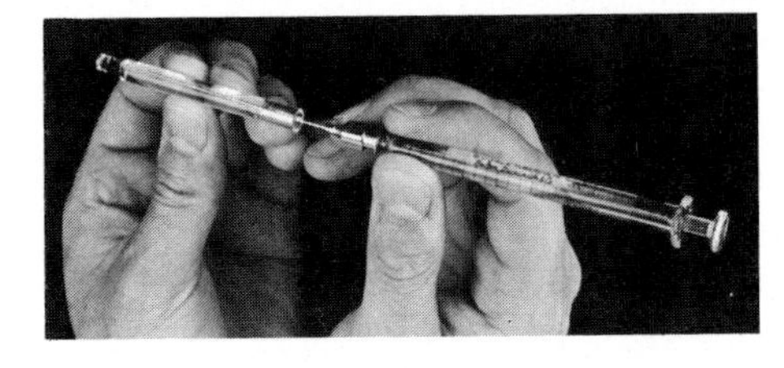

Fig. 26. – Removal of the polyacrylamide gel from the glass tube with an injection needle.

on the needle. Should difficulties arise, an attempt is made to advance toward the center from both tube ends and to detach the gel. The needle can be readily withdrawn from the gel tube if it is continuously rotated around its axis. The needle can also be replaced by a thin, but sufficiently strong, stainless steel wire which is pulled through the entire tube and rotated back and forth.

In some laboratories, the tube is dipped into a dish of cold water and the gel is ejected from the tube by compressed air. To prevent the gel from breaking, the pressure must be carefully regulated. WEINSTEIN and DOUGLAS [W 7] use a pipette control apparatus (available from firm 18) instead of compressed air, while SCHRAUWEN [S 23] loosens the gel from the inside wall of the tube with a sharp jet of water from an injection cannula. A special pressure gel remover is available from firm 5. In any case, excessive pressure, tension as well as scratching of the gel surface must be avoided; the latter would impair the densitometric evaluation.

In case of high percentage gels it might be feasible to incorporate a nonionic detergent (Triton X-100 or Brij 35) into the gels which helps release from the tube [N 27, P 11] (page 39). Silicon coated glass tubes may facilitate gel removal but are not recommended generally (s. page 58).

2.3.8.2. Detection Reagents and Color Reactions

2.3.8.2.1. General Color Reactions of Proteins

Of the dyes listed in Table 10, *Amido Black 10 B* is used most frequently. It should be pointed out, however, that the staining properties of Amido Black and other dyes may differ considerably depending on the batch and manufacturer [P 6]. For a quantitative densitometric evaluation, a calibration curve must be constructed for each dye batch and each particular protein.

Staining may be accelerated by incubating the gel at 96° C in 7 % acetic acid containing 1 % Amido Black for only 10 min. immediately after electrophoresis. This slows down free diffusion, particularly of weak bands (see page 81) and results in improved resolution [R 44]. Using electrophoretic destaining for 10 min. gels may be ready for viewing within less than $^1/_2$ hr.

The property of Amido Black to stain proteins metachromatically, resulting in stained bands of different shades of blue, black or even brown, is a drawback for quantitative densitometry. Johns [J 7a] found that the quantitative determination of histones requires an independent assay of the color yield for each band.

Light Green (C. I. 42,095) is not as favorable since its solutions are unstable and the color adsorption per protein quantity is not constant [B 43]. The same is true for *Nigrosin* (C. I. 50, 420), which is reported to have a high protein specifity [N 20] but gives rise to difficulties in the destaining step; for the dye contains components with a low mobility. Nigrosin, however, is reported [L 13] to stain wheat gluten proteins much more clearly and more permanently than Amido Black (0.0125 % in methanol — water – acetic acid 4:5:1, V/V). Ornstein [O 9] proposed *dibromo-trisulfofluorescein*; this dye binds cationic groups of acid-denatured proteins (λ_{max} = 525 nm). Montie et al. [M 62] stain proteins from Pasteurella pestis with a 2 % solution of *Ponceau* in 5 % trichloroacetic acid, while Dowding and Tarnoky [D 30] prefer an 0.6 % solution of Ponceau S (C. I. 27, 195) in 3 % trichloroacetic acid (destaining in 7 % acetic acid). Sastry and Virupaksha [S 5] add 0.01 % *aluminum lactate* to the Amido Black solution which is reported to enhance the color intensity.

Attention should be drawn to the dyes *Procion Brilliant Blue RS* and *Coomassie Brilliant Blue R 250* which were introduced by St. Groth et al. [S 74] for cellulose acetate electrophoresis (Table 10). The stains are distinguished by a high sensitivity and remarkable uniformity of protein binding, which is an advantage for densitometric protein quantitation.

Procion Brilliant Blue RS forms a covalent, nearly irreversible protein bond by the reaction of its chlorine-substituted triazinyl rest with hydroxyl, amino, amide and peptide groups of proteins. *Coomassie Brilliant Blue R 250*, on the other hand, in a weakly acid medium reacts with NH_3^+ and nonpolar protein groups. Coomassie is three times more sensitive than Procion. Azocarmine, Amido Black and bromophenol blue are $^1/_5$ to $^1/_{10}$ as sensitive as Coomassie. If a high protein sensitivity is desired, Coomassie is to be recommended, while Procion will be chosen for high demands of reproducibility and stable protein binding (densitometric evaluation) even for the nanogram range [G 41]. However Coomassie deviates notably from Beer's law at higher protein concentrations and therefore cannot be used to quantitate gels in which widely different amounts of proteins are present.

Table 10: Dyes and staining reactions of proteins.
Commercial sources for stains: for example, the firms 5, 10, 13, 20, 30.

Stain: Name and synonyms	Gross formula and molecular weight	λmax (min)	Fixing soln. and fixing time	Staining soln. and staining time	Remarks
Amido Black 10B, Naphthalene Black 10B, Naphthol Blue Black B, Buffalo Black (C.I. 20,470)	$C_{22}H_{16}O_4N_6S_2$ 572	620	Usually 1% (or less) in 7% acetic acid; 0.5–2 hrs. Also 1% in 7% acetic acid + 30% methanol (for intensified fixing)		Filter solution before use. On aging, dye precipitates and protein bands have black shade instead of deep blue color.
Procion Brilliant Blue RS (ICI), Reactive Blue 4 [S 74]	(Dichlorotriazinyl derivative of an anthraquinone dye)	602	20% aq. sulfosalicylic acid; 0.5–2 hrs.	1% in 10% acetic acid + 50% methanol; 1–2 hrs. Destaining by washing with solvent mixture	Always use fresh acid solutions for staining. Covalent dye-protein bond.
Coomassie Brilliant Blue R 250 (ICI), Supranol cyanin 6 B extra (Bayer), Acid Blue 83, (C.I. 42,660) [S 74]	$C_{42}H_{36}O_6N_3S_2$ 742 (triphenylmethane dye of magenta family)	590	20% aq. sulfosalicylic acid; 18 hrs.	0.25% in heavy metal free water; 0.5–2 hrs. [M 3]. Or: 0.25% in 9% acetic acid + 45% methanol; 30 min electrophoretic destaining in 3.75% acetic acid [M 46]	Electrostatic and van der Waals bonds between dye and protein.
Fast Green FCF, Food Green 3 (C.I. 42,053)	$C_{37}H_{31}N_2O_{10}S_3Na_2$ 808 Acid dye	625	1% (w/v) in 7% (v/v) acetic acid for 2 hrs. Destaining in 7% acetic acid by the leaching method		Suitable for quantitative densitometry [G 24a], light proof

According to Chrambach et al. [C 36], destaining even becomes unnecessary if the polyacrylamide gel is stained in a colloidal solution of *Coomassie* in 12.5 % *trichloroacetic acid* (TCA). The dye is relatively insoluble in the latter; consequently, it is selectively bound by protein bands. This staining method is outstanding for its rapidity and high sensitivity for protein detection in all gel systems; in this respect, it often surpasses Amido Black.

Immediately after electrophoresis, the gel is shaken for 30 min. at room temperature in 10–40 times the volume of a 12.5 % TCA solution. This is followed by incubation for 30–60 min. in a freshly prepared mixture of 1 vol. part 1 % aqueous Coomassie solution and 19 vol. parts 12.5 % TCA solution. Finally, the gel is transferred into 10 % TCA solution and photographed (red filter, e. g. R–2–25A, for densitometry Wratten No. 16 filter). If the gel contains 6–8 M urea, the 12.5 % TCA solution is replaced by a solution containing 5 % TCA and 5 % sulfosalicylic acid. Coomassie can be eluted from the protein bands by 90 % formic acid.

Koenig et al. [K 26a] stain overnight in a freshly prepared mixture of 1 part 1% aqueous Coomassie and 40 parts of 6% TCA in acetic acid-methanol-water (7 : 20 : 80, V/V/V) and destain with acetic acid–methanol–water (20 : 120 : 280, V/V/V).

Xylene Brilliant Cyanine G (*Coomassie Brilliant Blue G 250*, C. I. 42, 655, λ_{max} 610 nm, available from the firms 13, 30) is reported [S 4] to be three times more sensitive than Amido Black. It is slightly less sensitive than Coomassie R 250 but may be preferable to the latter since it does not form precipitates, is readily soluble in 7% acetic acid and does not fade during light exposure [M 31a]. For gel staining a 1 % solution in 7 % acetic acid can be used, or a 0.1 % solution in methanol-water-acetic acid (10:10:1, V/V) after protein fixation for 20 min. in 5 % TCA and several water washes.

Red W (available from firm 5) is a red general stain with about the same sensitivity as Amido Black. Red W can be used with other contrasting-colour specific stains (e.g. Sudan Black). Because it is transparent in red light it permits differential densitometry. By making two traces, one each with red (Wratten No. 42) and green (Wratten No. 57) light filters, one can eliminate anything such as lint or dirt (which absorb red light) as being non-protein.

Histone fractions can be stained specifically with components of Amido Black 10 B. Smith [S 56] found a red and a blue component when he chromatographed Buffalo Black NBR (Allied Chemical Co.) on Sephadex G–25: the red component exhibited no specificity, but the blue stained lysine-rich histones blue while other fractions only stained gray-blue.

Several methods are reported for staining of proteins in *sodium dodecyl sulfate* (*SDS*)-containing gels.

According to Vito and Santomé [V 6] the gels are first incubated for 4 hrs. in a mixture of equal parts of methanol and 20 % acetic acid, followed by 12 hrs. in a staining bath of equal parts of Amido Black solution in 10 % acetic acid and a mixture of 10 % acetic acid and 90 % methanol.

Maizel [M 3] prefers the following procedure: The gels are first soaked for 16 hrs. in 20 volumes of 20 % sulfosalicylic acid, then stained for 2–4 hrs. in 20 volumes of 0.25 % Coomassie Brilliant Blue and finally washed with 7 % acetic acid.

WEBER and OSBORN [W 5] use a staining solution prepared by dissolving 1.25 g of COOMASSIE in a mixture of 454 ml of 50 % methanol and 46 ml of glacial acetic acid, and removing insoluble material by filtration. Following electrophoresis the gels are stained at room temperature for 2–10 hrs. Destaining: after rinsing with water and incubating in destaining solution (75 ml of acetic acid, 50 ml of methanol and 875 ml of water) for at least 30 min., the gels are destained electrophoretically using the destaining solution.

DUNKER and RUECKERT [D 39] first leach out the SDS and precipitate the proteins by soaking the gels for 18–24 hrs. in 20 % sulfosalicylic acid. Next the gels are immersed for 4–6 hrs. in 0.02 % Coomassie freshly diluted in 12.5 % TCA (see page 74, [C 36]). After decanting the dye, 10 % TCA is added and the gels are allowed to stand in faintly blue solutions for several hrs. depending on the gel concentration (up to 48 hrs. for 15 % gels).

For the identification of *polypeptides* which are localized in a protein band, but differ in their chain ends, CATSIMPOOLAS [C 12] uses 1-fluoro-2,4-dinitrobenzene (FDNB) for staining.

Fluorescence quenching may be used to visualize proteins prior to cutting-up of gels for subsequent elution. PETERSON [P 15] uses a fluorophor-coated glass plate, illuminated with 254 nm radiation across the gel which is laid on a mask over the plate. In the main bands the fluorescence is quenched. This method was designed for use with fat gels (ø 2.2 cm) and simple protein mixes where a dominant band (trypsin) was sought. It might be less suitable with regular-sized gels containing complex protein patterns due to its limited sensitivity.

However the sensitivity of the *fluorescence quenching* method for protein detection can be greatly increased if a fluorescent dye such as N-(2-propenyl)-1-dimethylaminonaphthalene-5-sulfonamide (*N-allyl-DANSA*) is incorporated into the polymerizing polyacrylamide gel. Following electrophoresis the gel is stained with Amido Black, destained and scanned in a fluorometer. This technique, developed by BORRIS and ARONSON [B 48a], is about 50 times more sensitive than the common light absorption technique. The N-allyl-DANSA reagent can be easily prepared from DANSYL chloride and allylamine. The incorporation of the fluorescent into the gel does not affect the electrophoretic mobility of proteins compared with standard gels.

Direct fluorescent labeling of protein with anilinonaphthalene sulfonate (ANS) provides a simple method for rapid staining of proteins still retaining their enzymic or antigenic activity due to minimal denaturation [H 11]. Although less sensitive than dye-staining with e.g. Amido Black, the technique has been found useful for preparing small quantities of proteins to serve as antigens in the production of antibodies.

An aqueous stock solution of ANS (Mg salt, 1 mg/ml) is diluted with 0.1 N sodium phosphate buffer pH 6.8 to a final concentration of 0.003 % ANS. Freshly run disc gels are immersed into this solution for 3 min. and observed under a UV lamp. Denaturation of the proteins, prior to labeling, by immersion of the gels into 3N HCl up to 2 min. greatly enhances the intensity of staining. It is important to store ANS solutions in brown bottles with refrigeration since exposure to light greatly inactivates the stain. ANS is available from the firms 7a, 30.

1 - Ethyl - 2 - [3-(1-ethylnaphtho [1,2d] - thiazolin -2-ylidene) -2- methylpropenyl]-naphthol [1,2d] thiazoliumbromide not only stains RNA bluish-purple, DNA blue and protein red, but also acid polysaccharides, hence was called *Stains All* [D 1a]. For use a 0.1% solution in 100% formamide (pH 7.4) is diluted 1:20 to a final concentration of 50% formamide in water, the gel soaked overnight under light protection and rinsed free of excess stain with tap water. Due to the photosensitivity of the stain the gel should not be exposed to intense light. Available from the firms 7a,30.

2.3.8.2.2. Color Reactions of Metal-containing Proteins

Hemoglobin, myoglobin and other iron-containing proteins:

According to ORNSTEIN [O 8]: Sodium acetate (trihydrate) (16 g) is dissolved in 100 ml 7 % acetic acid, the solution is saturated with disodium ethylenediamine-tetraacetate (Na_2EDTA) (about 1 %), filtered, saturated with *benzidine hydrochloride* and again filtered. Of this stock solution, 10 ml are treated with 0.1 ml 3 % hydrogen peroxide. The gels are incubated in this mixture at 20° C until staining is sufficient (30–40 min.).

Ceruloplasmin is identified with the *dianisidine* color reaction [F 13a, O 13].

2.3.8.2.3. Color Reactions of Lipoproteins

(1) According to RESSLER et al. [R 31]: One volume part saturated *Sudan Black B* solution (in ethylene glycol) and 2 volume parts lipoprotein solution are allowed to stand at 3° C for at least 24 h. In the subsequent electrophoretic separation, the excess dye remains at the start. λ_{max} of dye absorption: 590 nm. See also [R 33].

(2) According to RAYMOND et al. [R 12]: Serum (200 µl) is shaken with 100 µl filtered 1% *Lipid Crimson* solution in diethylene glycol for 5 min. and centrifuged for 20 min. The reddish-purple supernatant is subjected to electrophoresis. Lipid Crimson (C. I. 26,105) is 4–o–tolylazo–o–toluidine–2–naphthol, m. p. 184–185° C. Suppliers: The firms 13, 30.

(3) According to NARAYAN and KUMMEROW [N 14]: *Iodine* (1–2 g) is dissolved in 150 ml glacial acetic acid, the solution is brought to 2 l with water and filtered. The gels are incubated with the solution. Lipoproteins which have already been detected with Sudan Black B can be fixed and counterstained permanently with the iodine solution. – The iodine solution is also suited as a general protein stain. Advantage: rapid staining without the need to remove excess stain; disadvantage: less sensitive and permanent than Amido Black 10 B which is a very sensitive stain, especially for high density lipoproteins. – NARAYAN and KUMMEROW [N 14] found that *Oil Red-O-trichloroacetic acid* offers no advantages compared to the Sudan Black B stain for lipoproteins.

(4) According to PRAT et al. [P 26a]: 500 mg Sudan Black B are dissolved in 20 ml acetone, then a mixture of 15 ml acetic acid and 80 ml water is added. After 30 min. of stirring the solution is centrifuged. Overnight staining is followed by destaining in three subsequent washes of a mixture of 150 ml acetic acid, 200 ml acetone and 650 ml water. The stain is stabile for about 2 days only.

2.3.8.2.4. Color Reactions of Glycoproteins and Mucopolysaccharides

(1) According to FELGENHAUER [F 13a] with *periodic acid-Schiff's* reagent (PAS):

Following electrophoresis, the gels are shaken overnight (16 hrs.) in a solution of 2.5 g sodiumperiodate, 86 ml H_2O, 10 ml glacial acetic acid, 2.5 ml concentr.

HCl and 1 g TCA. After washing for 8 hrs. with several changes of a solution of 10 ml glacial acetic acid, 1 g TCA and 90 ml H_2O, the gels are stained for 16 hrs. with Schiff's reagent and preserved by washing 2 times 2 hrs. in a solution of 1 g potassium bisulfite ($KHSO_3$), 20 ml concentr. HCl and 980 ml H_2O. All steps are done at 4° C with shaking. Densitometric recording at 543 nm. Subsequently the gels may be stained with Amido Black: 2 hrs. at 20° C in a solution of 0.5% Amido Black in 5% acetic acid, and destained in 2% acetic acid for 16 hrs. After recording at 626 nm visual comparison and graphical substraction of the two tracings allow the evaluation of carbohydrate content.

(2) According to Zacharius et al. [Z 1] with the PAS-reaction. See Table 11.

Table 11: Staining of Glycoproteins in Polyacrylamide Gel acc. to ZACHARIUS et al. [Z. 1].

Step	Gel treatment at 25° C	Time interval, min.
1	Immerse in 12.5 % trichloroacetic acid (25–50 ml/gel)	30
2	Rinse gently with distilled water	0.25
3	Immerse in 1 % periodic acid (made in 3 % acetic acid)	50
4	Wash 6 × for 10 min. each in 200 ml distilled water/gel with stirring or shaking or wash overnight with a few changes	60 or ON[b]
	If 60 min. washing was used, check last wash with 0.1 N $AgNO_3$: and when test is negative for periodate continue washing with 2 more changes	20
5	Immerse in fuchsin-sulfite stain in dark[a]	50
6	Wash with freshly prepared 0.5% metabisulfite 3 × for 10 min. each (25–50 ml/gel)	30
7	Wash in distilled water with frequent changes and motion until excess stain is removed	ON
8	Store in 3–7.5 % acetic acid	

[a] Prepared according to MCGUCKIN and MCKENZIE [M 38].
[b] Overnight.

(3) Other procedures on the basis of the PAS-reaction were reported by CLARKE [C 41], CALDWELL and PIGMAN [C 4] and KEYSER [K 13].

(4) *Acid glycoproteins* can be stained by incubation for 4 hrs. in 0.2 % *Alcian Blue* solution (C. I. 74,240) according to CALDWELL and PIGMAN [C 4]. Destaining takes place by electrophoresis in 7.5 % acetic acid (15 mA/gel).

(5) *Acid mucopolysaccharides* can be localized by staining for 1 hr. in 0.1 % *Toluidine Blue 0* solution in 1 % acetic acid to a red color. Destaining is carried out by leaching the excess dye in 1 % acetic acid. RENNERT [R 28] uses a 1 % solution in 3.5 % acetic acid.

The fact that the PAS reaction is sensitive to glycoproteins *and* acid mucopolysaccharides, while Toluidine Blue reacts with acid mucopolysaccharides *and* nucleoproteins, makes it possible to differentiate between glycoproteins and mucopolysaccharides; if both reactions are positive for the same substance, an acid mucopolysaccharide is present.

(6) Using *Amido Black* rather than PAS for orosomucoid, KROTOSKI and WEIMER [K 36] demonstrated better resolution, when the stain is made up in a 50 % methanol rather than in water. This gives improved fixation of the glycoprotein, but results in same shrinkage of the gel.

2.3.8.2.5. Color Reactions of Specific Proteins

Specific detection methods have been reported for haptoglobins [F 13a] and tryptophane- and tyrosine containing proteins [F 14a].

2.3.8.2.6. Color Reactions of Deoxyribonucleic Acid (DNA)

(1) Deoxyribonucleic acid (DNA) can be distinguished from ribonucleic acid (RNA) by the *diphenylamine* reaction:

The gels are stored in a mixture of 10 ml 1 % diphenylamine solution (in glacial acetic acid) and 1 ml 10 % sulfuric acid for 1 hr. Subsequent immersion in boiling water for 10 min. results in diffusely blue bands, the color intensities of which fade during storage in 33 % acetic acid.

(2) Feulgen stain for DNA:

After incubation for 30 min. in ice-cold 1 N HCl (20 ml/gel), the gel is rapidly transferred into 1 N HCl at 60° C and allowed to stand for 12 min. Subsequently, it is treated with Schiff's reagent for 1 hr. at 20° C. Storage takes place in Schiff's reagent (1 month) or in an 0.25 % solution of potassium bisulfite in 0.05 N HCl ($>$ 1 month).

(3) *Native DNA* can be identified with *Methyl Green* (C. I. 42,590) according to KURNICK [K 40]:

Prior to the color reaction, the 0.25 % solution of the stain in 0.2 M acetate buffer of pH 4.1 should be repeatedly extracted with chloroform until the chloroform phase is no longer stained (by Gentian Violet). The solution is stored in darkness in a brown bottle. It can be used repeatedly, provided that the pH value does not drop below 4.5. Incubation proceeds for 1 hr. at room temperature and excess stain is leached out by repeated replacement of the washing liquid (0.2 M acetate buffer, pH 4.0, 1–3 days).

(4) *Denatured DNA* can be identified with *Pyronine B* (C. I. 45,010) according to KURNICK [K 41] as modified by BOYD and MITCHELL [B 50]:

Prior to use the 2% aqueous stain solution should be purified by exhaustive extraction with chloroform. The aqueous phase is stored at + 4° C in darkness. Before use, it is diluted to the 10-fold volume with 0.2 M acetate buffer, pH 4.5. After use, the dilute solution is discarded.

For the identification of denatured DNA, the gels must first be kept in 0.1 M acetate buffer, pH 4.5, for 1 hr. or more. They are then incubated for 1 day in the dilute Pyronine B solution. The stain excess is removed by leaching with the acetate buffer.

Oligodeoxynucleotides are stained with *Toluidine blue O* (E 4):

The gels are kept overnight in a 1 % acetic acid solution of 0.01 % (w/v) Toluidine blue 0. Excess dye is removed by extended washing periods with large volumes of 1 % acetic acid.

2.3.8.2.7. Color Reactions of Ribonucleic Acids (RNA)

1. RICHARDS et al. [R 34] describe the following mixture as a specific reagent for RNA (particularly those of lower molecular weight, such as t-RNA):

1 % Lanthanum acetate and 2 % *Acridine Orange* (C. L. 46,005) are dissolved in 15 % acetic acid. Staining takes place for 12 hrs. (overnight) and destaining by exhaustive leaching with water or by electrophoresis (p. 80); lanthanum acetate precipitates and binds the RNA in the gel; Acridine Orange stains the precipitate in orange red.

The *Acridine Orange* bond depends on the base composition of the RNA, however [B 16]; it therefore differs from one RNA to another.

2. GROSSBACH [G 43] prefers *gallocyanine-chrome alum* as an RNA stain, because it reacts with the phosphate groups of the nucleic acids and consequently also furnishes uniform dye complexes with different types of RNA. In this case, the RNA-dye complex follows the Lambert-Beer law within certain concentration ranges.

After electrophoresis, the gel is first fixed for 12 hrs. in a solution of 1 % lanthanum acetate in 15 % acetic acid. This is followed by staining for 24 hrs. in a freshly prepared gallocyanine-chrome alum solution. Destaining takes place by leaching the dye excess out with the fixing solution. – Preparation of the gallocyanine-chrome alum solution [D 11]: gallocyanine (600 mg) is shaken in 200 ml water for 1 min. and filtered; the filtrate is discarded. The filter paper with its residue is then boiled in 200 ml of a 5 % aqueous chrome alum solution on a water bath for 30 min. After cooling, it is filtered; the filtrate is adjusted to pH 1.6 with 1 % hydrochloric acid. Nucleic acids assume a deep blue color.

3. GROSSBACH [G 43] reports that the electrophoretic migration of RNAs can be observed if they are stained with 0.05 % *Pyronine G* before electrophoresis; their separation behavior is not modified. Free Pyronine migrates with the buffer front during electrophoresis.

4. PEACOCK and DINGMAN [P 8] first incubate the RNA containing gels in 1 M acetic acid for 15 min. and then stain for 1–16 hrs. with an 0.2 % solution of *Methylene blue* in a mixture of equal parts of 0.4 M sodium acetate and 0.4 M acetic acid (pH 4.7); they destain by washing with water. The stain, however, is not quantitative: small RNA may stain more intensively than large RNA.

5. KONINGS and BLOEMENDAL [K 30] stain for 4 hrs. with a fixative stain solution containing 0.2 % *Toluidine blue 0* in 10 % acetic acid and destain electrophoretically in 2 % acetic acid by passing 20 mA per gel during 20 min.

6. DAHLBERG et al. [D 1a] uses Stains-All (see page 76).

2.3.8.3. Staining

Staining takes place by incubating the gels with the staining solutions in test tubes or in a staining tank (Fig. 27).

The staining tank contains the same number of compartments as the number of gels which can be analyzed in the electrophoresis apparatus. If the bottom of the compartments is perforated, the tank can be used advantageously both for staining and for destaining by the diffusion method (p. 81). – SCHRAUWEN [S 23] uses so-called "pickle" tubes in which the gels are wetted uniformly from all sides. These tubes are particularly suited for incubation with enzyme substrates.

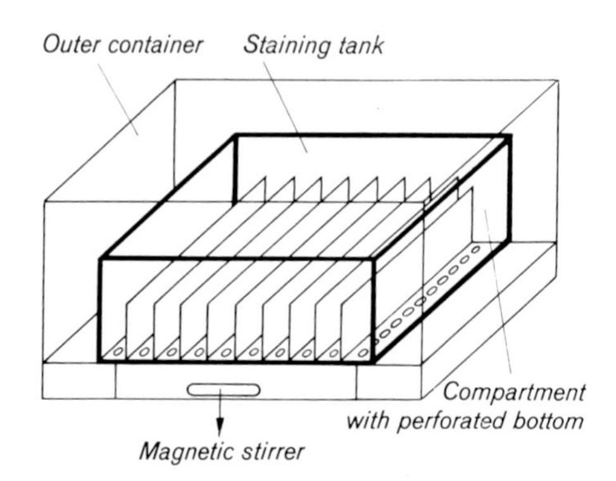

Fig. 27. – Staining and destaining tank (for destaining by the leaching method).

The *staining period* for most protein stains like Amido Black amounts to about 0.5–2 hrs. at room temperature, depending on the gel dimensions (thickness), concentration, and the quantity and molecular weight of the separated components. Since the progress of fixing and staining within the gel is subject to diffusion laws, allowance should be made for sufficient fixing and staining periods. The solutions should be frequently agitated. It must be noted that the dye concentration of the staining solution decreases with repeated use, depending on the volume ratio of gel to staining solution. See also page 72 for staining at elevated temperatures.

2.3.8.4. Destaining

Destaining can be carried out by exhaustive leaching of the excess of stain, which is the mildest but most time consuming method, or by electrophoresis of the unbound stain. The advantages and disadvantages of the two methods are listed in Table 12.

The choice of the destaining method depends essentially on the properties of the separated substances. If they can be easily fixed the *electrophoretic method* will be preferred; it furnishes satisfactory results in most cases. A spacer gel plug of 0.5 cm height is first polymerized over the lower opening of the destaining tube. The stained gel is then inserted into the tube, is covered by destaining solution to the top of the stained gel and then with acetic acid

Table 12: Comparison of destaining methods. From: MAURER [M 28].

Method	Advantages	Disadvantages
Simple leaching (by diffusion)	Weakly bound proteins are preserved	Time-consuming; background cannot always be eliminated, thus making densitometric analysis difficult
Electrophoresis of unbound dye	Rapid, complete destaining (20–40 min.)	Loss of weakly bound proteins; gradual fading of color intensity and rise of electrical resistance during electrophoresis possible due to gas bubble formation

up to the upper tube rim. The destaining solution is prepared by photopolymerizing an aqueous solution containing 6 g acrylamide, 0.5 mg riboflavin, 50 μl TEMED per 100 ml for 1.5 hrs. and diluting this solution with an equal volume of distilled water. The tube is inserted into the electrophoresis apparatus, the two electrode reservoirs are filled with 7 % acetic acid and about 30 μl stain solution (e.g 1 % Amido Black solution) is added to the cathode solution (cathode at top; for reason, see p. 82). Finally the electrodes are connected with the power pack. In many cases, a current of 10 mA/gel is sufficient to obtain complete destaining in 40–60 min. For other destaining equipment, see p. 39.

It should be noted that some laboratories strictly adhere to the *leaching method* for destaining so as not to loose weakly bound quantities of protein. That this danger actually exists was demonstrated by several investigators [B 20, J 12, M 20, N 3]. Thus, after a customary electrophoretic destaining procedure, certain histone fractions were absent which were retained when the leaching method was used [J 12]. Moreover, the migration of stained proteins was observed during electrophoretic destaining [B 20, M 20, P 24].

Systematic studies [M 31] on basic proteins of relatively low molecular weight (histones) showed that at least 2.5 hrs. are necessary for sufficient staining in 1 % Amido Black solution (20 % ethanol, 7 % acetic acid). This may be accelerated by heating to 95° C for 10–15 min. Destaining by the first method in these cases is successful only after a long period of time (Table 12); frequently, the blue background due to stain decomposition cannot be eliminated, thus leading to difficulties in the densitometric evaluation. In these cases, a mild electrophoretic method is advisable; *destaining* takes place *perpendicular to the direction of separation* of the substances. This reduces the destaining time and decreases the required voltage gradient.

2.3.8.5. Comments on Staining and Destaining

The efficiency of fixing and staining of the separated components as well as of destaining determines the quality of the quantitative evaluation. A few brief remarks will define the problem. One of the most important factors concerns the ratio between the quantities of dye and sample molecules (e.g. protein). For quantitative evaluation the ration should always be constant. This may be true for one molecule within a certain limited range, but not for molecules of different size, shape and charge [S 76].

Thus, protein quantities of the same type (e.g. albumins) will adsorb proportional quantities of dye within certain concentration ranges, while the same quantities of proteins of different types (albumins and globulins) bind different quantities of dye. Consequently, correction factors must be calculated with which the percentages of e.g. the total proteins determined by densitometry are corrected. Naturally, when identical protein bands of different samples are to be compared and quantitated, i.e. if only relative values are to be determined, such factors are unnecessary. On the other hand, if the binding capacity of different dyes is compared for the same sample molecules, the ranges for which a proportionality between the quantities of dye and molecule exists, will be found to differ (compare Procion Brilliant Blue RS and Coomassie Brilliant Blue R 250, p. 72). In particular, the binding power differs considerably in the case of high sample concentrations.

Another aspect concerns the reversibility of the sample-dye bond. Although the reaction equilibrium between sample and dye is shifted in favor of the bond, retention of the equilibrium requires a small quantity of free unbound dye [D 7]. If the equilibrium is disturbed by excessive electrophoretic destaining, the color intensity may fade; this is prevented by the addition of very small quantities of dye to the destaining solution. Moreover, it is important whether precipitation and denaturing of the separated molecules are complete within the entire depth of the gel. If this does not occur, artifacts of the most diverse nature can be expected; for example, weakly bound proteins can continue to migrate during electrophoretic destaining (see above); moreover, the dye affinity of the proteins is modified.

2.3.9. Quantitative Evaluation

Since polyacrylamide gels are transparent, the component patterns can be quantitated by optical methods: either by direct densitometry of the gels or by densitometry of photographs.

2.3.9.1. Direct Densitometric Analysis

Ordinary densitometers designed for flat media (e.g. paper strips) do not possess optical characteristics or resolving capacities suited to the cylindrical shape and fine band structure found in disc gels. However, with densitometers designed for use with disc gels, cylindrical gels can be scanned with high resolution. In designing such densitometers three factors have to be considered [K 24]: (1) size of light beam, (2) scanning speed and (3) alignment (positioning) of the gel. The *light beam* should be narrow (less than 50 μm), yet intense enough to be registered, the *scanning speed* should be a compromise between rapidity and accuracy and will be limited by the mechanical inertia of the scanner and recorder, and last not least the *alignment* should be accurately yet easily manageable. Furthermore, a limiting factor in the resolution of adjacent peaks may be the focusing of the image of the slit in the light beam [L 36]. Owing to shortening of the focal length of the lens, the correct adjustment for this requires special attention.

2.3.9.1.1. Scanning of Stained Patterns and Problems of Quantitative Densitometry

Densitometers for disc electrophoresis must be capable to resolve protein bands of at least 50 μm thickness, with spaces at least 50 μm between them, and to furnish a sufficiently precise tracing at the wavelength of the absorption peak of the protein-dye complex (Fig. 28). In practice, most densitometers utilize the principle to move the gel at a constant speed perpendicular to a fixed parallel light beam.

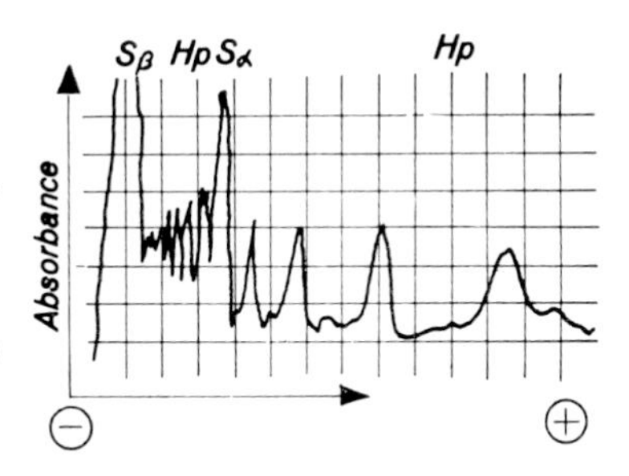

Fig. 28. – Portion of the tracing of a disc electropherogram (haptoglobin type 2-2-serum) by direct microdensitometry.

The haptoglobin (Hp) bands between Sβ (β_1-lipoprotein) and Sα (α_2-macroglobulin) have a spacing of 30–50 μ. Instrument: Model E of the firm 5.

It should be mentioned that in addition to the factors discussed on page 82, the following influences may cause a nonlinearity in the ratio between the light absorption of the sample-dye complex and the sample quantity: polychromism of the stain, interference and reflection as well as different band curvatures. These are partly counteracted by suitable filters and by a correct alignment and positioning of the gel: The gel is placed between two glass slides perpendicular to the path of the beam; in no case, however, should the gel be compressed to such degree that band curvatures result from the cylindrical shape of the gel. Cylindrical glass tubes can be used in well focused densitometers with only slight loss in resolution, provided the light beam height (corresponding to the length of the cross-sectional rectangle) is reduced considerably. Spherical aberration which may result as an optical error from the lens shape of the cylindrical glass [A 37] is mostly avoided by the use of 7 % acetic acid as liquid gel environment since both the gel and the acid show almost identical refractive indices.

Although the quantitative evaluation of disc electropherograms is of considerable interest, statistically based data on these problems are sparse. NARAYAN et al. [N 17] examined the quantitative reproducibility of several methods of sample application (see page 69). Overall reproducibility of integrated area values was within a standard deviation (s. d.) of 1 % for the relative percentage of the main component and of 0.7–3 % for the peak area of the main component. Amido Black stained gels stored over 3 months in acetic acid lost 15–25 % in overall absorption but relative values among peaks changed little. DIETZ [D 16] tested the reproducibility of the Model E Microdensitometer from firm 5 and found a s.d. of 0.61–1.81 % in the integrated areas under LDH isozyme peaks when the same gel was scanned repeatedly. When the same sample was electrophoresed and stained repeatedly, s.d. ranged from 0.37 to 2.18 %. FAMBROUGH et al. [F 6] found that binding of Amido Black to histones is linear up to 10 μg per fraction (s. d. ± 4 %). When the concentrations of fractions eluted from preparative disc columns, as measured by turbidity at 400 nm after TCA precipitation, were compared with those stained by Amido Black in analytical gels and traced by the Model E Microdensitometer, they agreed within 1 %.

KRUSKI and NARAYAN [K 37] found that the total peak area of the Amido Black-ovalbumin complex is proportional with increasing amount of only up to 15 μg protein in 7.5 % gels. More important, their data pointed out that, for the same amount, peak areas are greatly influenced by the concentration of the separation gel as well as by the migration distances. Peak areas became larger with increasing migration distances and gel concentrations. Therefore it is necessary to let the proteins migrate to fixed distances into the gel. For biological mixtures of proteins, it is not possible to present either the sample size or the migration distance of all the components to their optimum range. Large amounts of components and fractions of high molecular weight that do not penetrate sufficiently will appear to be present in smaller proportion in the mixture than they actually are. Kruski and Narayan therefore stress the difficulties in interpreting the results from densitometer traces.

GOROVSKY et al. [G 24a] reported that the acid dye *Fast Green* does not show the drawbacks of Amido Black and Coomassie R (page 72), yet has a similar staining sensitivity as Amido Black, hence may be particularly recommended for quantitative densitometry. A satisfactory, highly reproducible linear relationship was found between peak area and protein quantity in the range between about 5 and 150 μg (Fig. 29). See Table 10 for use of the dye.

2.3.9.1.2. Recording of Unstained Patterns by Direct Ultraviolet Absorbance Measurement

With increased use of a variety of stains which absorb at different wave lengths and with growing interest in direct UV measurements of proteins [D 34] and nucleic acids [L 35–36], efforts have been made to modify laboratory *spectrophotometers* for this purpose [T 1, D 34, G 36, K 24].

DRAVID et al. [D 34] compared scanning at 280 nm of unstained gels, contained in quartz tubes, with scanning at 550 nm of Amido Black stained gels and with tracing of photographs of stained gels (Fig. 30). Although similar resolutions are found in general, the relative proportions of some protein bands are changed. This may be contributed to two facts: (1) the differing protein binding capacity of Amido

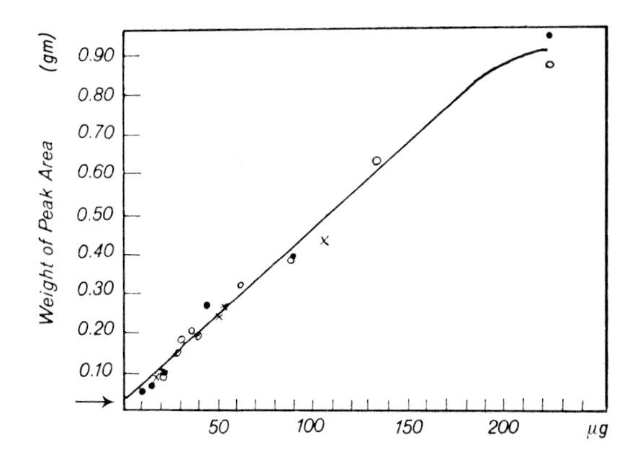

Fig. 29. Plot of peak area weight under densitometric tracings versus amount of protein placed on the gel.

Tube i. d. 6 mm, gel system No. 1 modified with 8 M urea, flagellar protein fractions from ● whole flagella, o microtubules, x matrix. Fast green staining (Table 10). The arrow indicates the level of staining and/or scatter obtained at the origin of a blank gel. From GOROVSKY et al. [G 24a].

Black and (2) the differing content in proteins of tryptophane and tyrosine residues mainly influencing the absorption at 280 nm. The gels cannot be scanned at 230 nm because of the high self-absorption of polyacrylamide gel, even if they are prepared from highly purified monomers.

Similarly WATKIN and MILLER [W 4a] scanned gels in quartz tubes at 280 nm (Fig. 31) in a special carriage which allowed to measure protein quantities between 1 and 200 µg in any single peak. Data on reproducibility, proportionality, accuracy and sensitivity of the method were given.

In principle, if the absorbance coefficient ε of a protein is known its amount may be calculated from the absorbance unit (AU) obtained by multiplying the band volume by the average optical band density. Since the latter is proportional to the peak area of the scan, multiplication of the peak area by an appropriate conversion factor gives the AU directly. In practice, it may be assumed that 1 mg of an average protein in 1 ml gives 1 AU at 280 nm in a 1 cm path cuvet. An accurate determination however must be based upon the true ε or $E^{1\%}_{280}$ value of any particular protein.

Some problems arising with UV scanning should be mentioned. Due to residual dust particles in the gel or scratches on the tube newly polymerized gels may show a baseline scatter between 0.01 and 0.02 OD within the 0–1 OD range. Sensitivity and accuracy of the method are dependent on the linearity of this baseline, however. – In gels of gel system No. 1 (but not No. 7), a large UV-absorbing peak moves behind the dye front (peak IF, Fig. 31), probably owing to unpolymerized and uncross-linked acrylamide which at the running pH of 9.5 can react as an acid. The peak area is relatively constant and apparently independent of the concentration of gel components. Preelectrophoresis should separate this peak from any fast moving protein. Moreover, the gel may occassionally shrink away from the tube

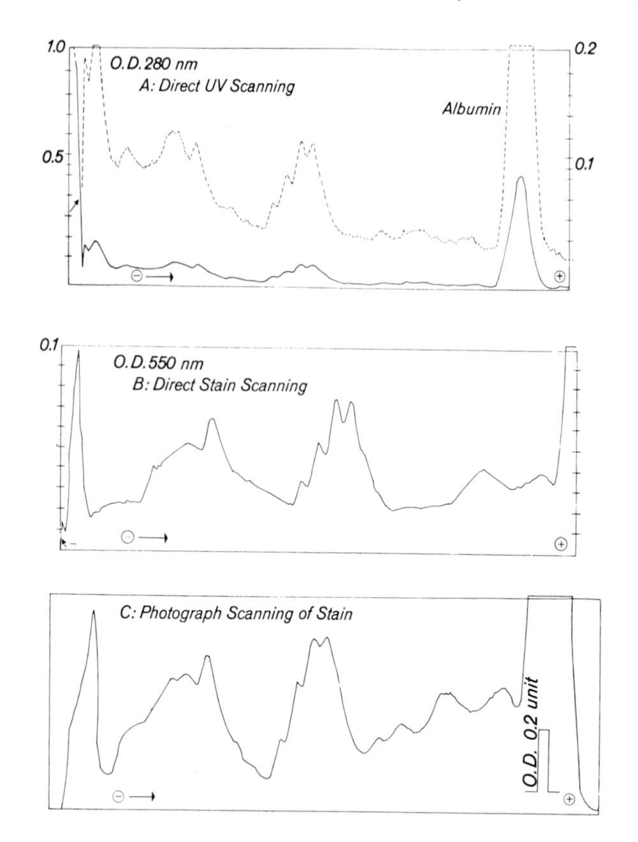

Fig. 30: Scans of disc electropherograms of calf serum proteins in 7.5 % gels. From: DRAVID et al. [D 34].

A: The unfixed and unstained gel scanned at 280 nm immediately after electrophoresis. Continuous line: full length of recorder chart set at 1.0 optical density (OD) unit, broken line: at 0.2 OD $_{280}$. Arrow: beginning of separation gel. B: Same gel, stained with Amido Black and scanned at 550 nm at 1.0 OD unit. C: Same gel as in B, photographed and its negative scanned. Recorder speed 10 × that of the gel throughout.

walls near the top in the absence of a spacer gel, resulting in an artificial peak (D, Fig. 31) near the water-gel interface.

2.3.9.1.3. Special Densitometers

As with modification of electrophoresis densitometers designed originally for low resolution tasks such as paper or cellulose acetate electrophoresis, the mere addition of narrow slits and extra signal amplification is not enough to achieve the high resolving capacity needed for the fine structure to be found in complex patterns like RNA or ribosomal, serum or bacterial proteins. Proper optical characteristics, including illumination of the sample in parallel light and subsequent magnification of the image prior to 25 μm wide slits are required.

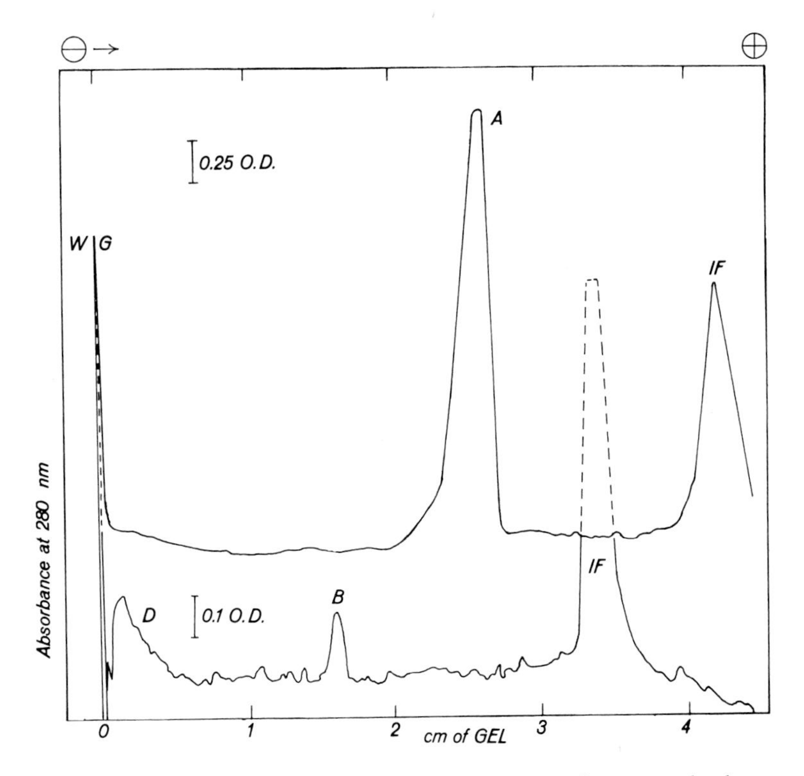

Fig. 31. Direct densitometric UV scanning at 280 nm of quartz gel tubes.
Tube i. d. 5 mm, gel system No. 1a with sodium sulfite added to 5 mM final concentration. A: 200 μg pepsin (0-2.5 OD range), B: 5 μg pepsin (0-1 OD range). WG: Water-gel interface, IF: impurity front, D: gel distortion From: WATKIN and MILLER [W 4a].

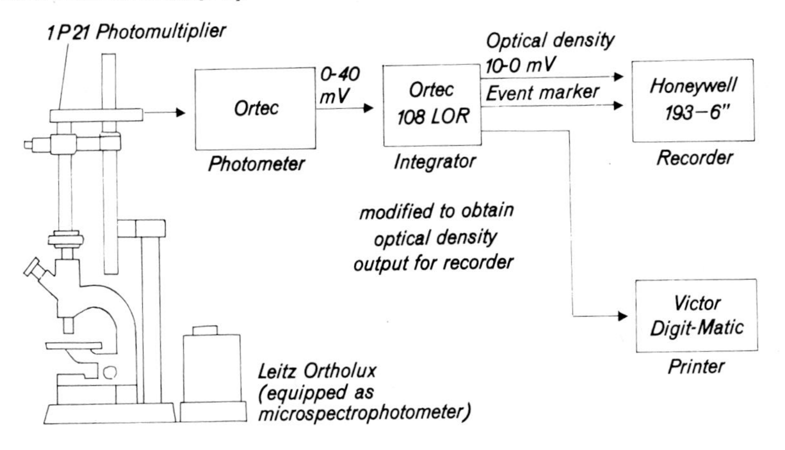

Fig. 32. – Circuit diagram of an automatic microdensitometer for a quantitative analysis with integrator and recorder. From: ALLEN and JAMIESON [A 19].

Excellent resolution and recording may be also achieved by *ultra-micro-densitometers* equipped with suitable optics. Such an equipment may consist of a microscope (e.g. Leitz Ortholux), photomultiplier, photometer, integrator, recorder and printer [A 19] as shown in Fig. 32. A similar ultra-micro-densitometer was employed by Hydén et al. [H 49] to analyze microgels of 0.2 mm diameter. Using a high duty amplifier, the authors obtained a clear recording even of weak, barely visible protein bands.

Ultra-microdensitometers of this or similar type are commercially available e.g. from the firms 5, 17 and 22. Modified low resolution microdensitometers suitable for relatively simple disc patterns such as hemoglobin, isoenzymes or well separated large protein bands wider than about 250 μm are abailable e.g. from the firms 2, 9, 17, 25 and others. Spectrophotometer attachments fulfilling similar requirements but delivering monochromatic light are manufactured by the firms 11, 12 and 35. High resolution (15–20 μm) UV and visible light recording microdensitometers, specifically designed for disc cylindrical gels, are available from the firms 5 and 35.

Petrakis [P 16] constructed an inexpensive densitometer capable to resolve 15 μm wide bands spaced 15 μm apart.

Hannig and Wirt [H 7a, W 22] designed an *electronic densitometer* on the principle of the *flying spot* [O 9], thus avoiding errors by the normal optical split completely.

The image of the electrophoretic pattern is projected on a Vidicon camera using monochromatic light (visible or UV). The generated electrical output signal is proportional to transmission. It is converted to the logarithm by an analog computer and integrated. By normalizing the integral to 100 %, the integrals corresponding to the individual fractions can be used to determine the concentrations. This system offers the following advantages: First, any electrophoretic patterns can be analyzed, since images of approximately identical size are always projected; secondly, the shape of the bands is immaterial (for example, it may be curved), since the line scanning – line perpendicular to the direction of migration – allows each part of the stain to contribute to the integral.

2.3.9.2. Photographic-densitometric Analysis

The gels can be photographed under obliquely incident light. To obtain reproducible photographs, the use of a fixed system of microscope lamp, mounting plate with white cardboard as a reflector and camera is recommended [R 19, S 53]. A uniformly illuminated frosted glass plate which furnishes diffuse, homogeneously distributed light is even better. The gels are photographed in glass cells filled with 7 % acetic acid. Such photographs are sufficient for many analytical purposes. However, they are only suited for a quantitative densitometric analysis if the linear absorption range of the gel band pattern lays within the linear response of the photographic film [F 16] (Fig. 33).

This linearity is determined by the variation of absorbance along the picture of the optical wedge on the same photograph. Thus a set of gels is photographed in transmitted light with a linear optical wedge (e.g. 0.087 or 0.13 absorbance units

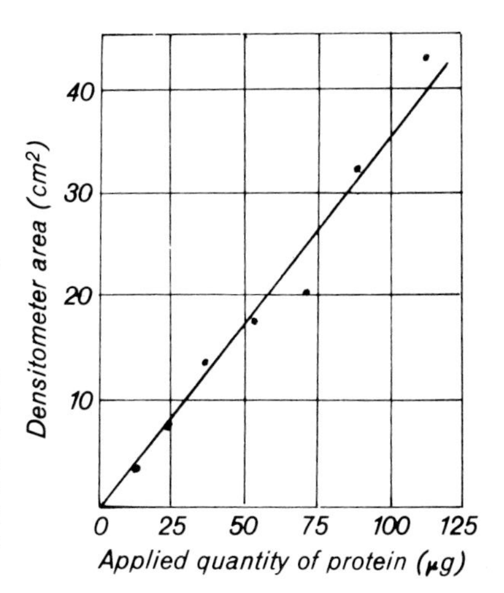

Fig. 33: Relation between applied quantity of protein and area under the densitometer curve of the gel photograph. From: FESSLER and BAILEY [F 16].

Amido Black stain of equal parts of α-and β-gelatines. Gels photographed in transmitted light with a linear optical wedge. Densitometry of the transparencies with Mark 3 C densitometer of firm 17. Selection of photographs by the absorption range of the gel band pattern which layed within the linear response of the photographic film. Area determination under the densitometer curves by planimetering after the determination of the base lines and divisions between incompletely resolved bands.

per cm). Photographic exposure time and time of destaining of the gels are varied. The transparent photographs are scanned in a densitometer and those photographs selected in which the absorbance range of the band pattern of the gels coincides with the linear range of the photographic film. – However, the usefulness of a photographic plate depends not only on its gradation (given by the slope of the linear segment of the characteristic curve) but also on its spectral sensitivity distribution and on the grain and total sensitivity. The gradation can be extended by prolonging the processing time and by other steps, so that brightness differences which are no longer visibly perceptible can still be measured.

In any case, the geometry of the measuring system also needs to be calibrated for a photometry of the films, since during transmission through a processed photographic emulsion light is lost by absorption and scattering [B 24]. Consequently, stray light must be taken into account in an exact quantitative evaluation. In micro-disc electrophoresis, photographic analysis appears to be the method of choice. The photographic enlargement can be conveniently analyzed with commercial densitometers.

In practice, one photographic film is used for the image of the investigated gel, a blank gel (without sample but treated in the same manner as the working gel) and the image of the optical wedge. All three of the images are photometered. The densitometer curve of the control gel serves as the base line.

This method is based on the assumption that the photographic resolving power is equal to that of the eye: two closely spaced bands which are observed to be closely adjacent to each other should also appear to be clearly resolved in the photograph. This cannot always be true for the following reasons [B 62]:

Fig. 34 is a schematic diagram of the set up for photographing a horizontally stretched gel of 40 mm length and 5 mm thickness. Trigonometric treatment shows that $\text{tg}\ \alpha = 20/160 = x/5$, and therefore $x = 5/8 = 0.625$ mm. With other

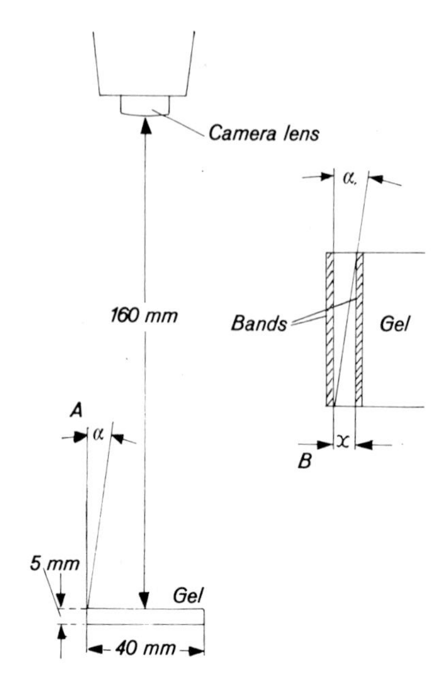

Fig. 34 : – Photographic analysis. According to BURNS and POLLACK [B 62].

(A) Schematic diagram of the photographing system for stretched gel. (B) Detail of A. Explanation below. Technical data: Polaroid camera (46-L projection film) with 6 + close-up lenses. Distance between gel and lens 15.5 cm, i.e. the radius of the arc in which the gel is located also amounts to 15.5 cm. The white background is illuminated with 2 floodlights from about 15 cm below. the gel; the distance of the lights from the center of the background is 30 cm for each (angle to the horizontal 30°). The photographs are enlarged and photometered.

words, if the separated protein bands are parallel, as is the case with the present arrangement, the bands which are observed with a spacing of less than 0.625 mm at the gel ends will no longer be resolvable by photography. Since band spacings of < 0.625 mm are by no means rare in disc gels, an identification may not be possible under these conditions. To obtain complete resolution, the smallest distance between two bands must be equal to $\alpha = LW/2\,d$ [mm] (L = length of gel, W = width of gel, d = distance of gel-lens). Consequently, the length and width of the gel should be reduced or the distance of the gel from the lens should be increased, which is not practically feasible. The problem can be solved by moving gel and lens towards each other so that the lens will "see" each part of the gel once. Most densitometers (page 88) operate by this principle. BURNS and POLLACK [B 62] have suggested another solution: They *bend* the gel such that the distances of all parts of the gel to the lens are identical.

A commercial gel-photographing device is available from firm 5.

2.3.10. Slicing of the Gel

The determination of radioactivity, enzyme, antigen and other activities in the gel frequently requires the gel to be cut into sections and determinations of the activities in the fractions.

A simple method consists of placing the gel between two combs and slicing them at the spacings of the teeth with a razor blade which has previously been dipped into a detergent solution [M 3]. In another procedure, the gel is inserted into a tube until the desired thickness of a section projects from the tube rim and can be cut off with a sharp blade. But how can the gel thickness be changed that it can be easily inserted into the tube? Polyacrylamide gels shrink in ethanol and swell in water; by incubating them in suitable mixtures, appropriate gel dimensions can be obtained. However, manuel slicing cannot yield sections of uniform thickness. Therefore, manual methods are generally not recommended.

Gel cutting devices producing uniform sections may utilize *three different methods*: (1) cutting with a number of fixed wires or blades (egg slicer principle), (2) moving the gel across a chopping blade (guillotine type), and (3) extruding the gel through a narrow orifice.

Cutting with closely spaced (~ 1 mm) small-gauge (~ 0.1 mm) stainless steel *wires* was applied in "*egg slicer*"-like constructions [C 35, H 16, L 24].

To slice gel cylinders *longitudinally* FAIRBANKS et al. [F 1] constructed the wire-device shown in Fig. 40 which furnishes gel strips suitable for autoradiography.

However, sharp *blades* are superior to wires for cutting low or high concentrated or frozen gels [G 20]. Devices with closely spaced blades (up to 50) were developed [A 37, C 2, C 34, G 20]. GOLDBERGER [G 20] utilizes rotating circular blades which provide a minimum pressure upon the gel. Consequently there is little tendency for the gel to be deformed. Although gel sections of satisfactory uniform thickness are obtained, the applicability of these devices is limited because of their fixed blades which cannot permit any desired slice thickness.

In principle, *freeze microtomes* may also be used for gel slicing; but usually these produce a slice thickness of only 40 µm or less of moderate or unsatisfactory uniformity. Moreover they are not equipped to handle gel cylinders as specimen. Nevertheless short gel pieces may be cut and several slices pooled to one fraction. It is advisable to cool the microtome on all sides in a cryostat. Due to ice crystal formation in the gel the regularity of gel slicing by usual microtomes is not satisfactory. The addition of sodium dodecylsulfate [B 31], glycerol [W 6a] or urea [G 46] to the gel buffer has been recommended to prevent that. But these agents lower the freezing point of the gel considerably. Replacing the gel water by paraffin however yields gels which can be fractionated to slices of any desired thickness with sufficient accuracy. GRAY and STEFFENSEN [G 32] dehydrated and embedded the gels in paraffin which they could uniformly slice into fraction thicknesses of 250, 125, 62.5 or less than 50 µm with a rotary microtome.

15 %-gels are left for 30 min. in 70 % ethanol, 60 min. in 70 % ethanol, 120 min. in 95 % ethanol, 30 min. in 100 % tert-butanol, 210 min. in 100 % tert-butanol, 180 min. in 100 % tert-butanol: paraffin (1 : 1) and at least 60 min. in paraffin, heated to the melting point. For counting of radioactivity see p. 98.

However, since the usual range of the desired slice thickness lies between about 0.1 and 2 mm and for the reasons given above, microtomes have not been favored for gel cutting in general. Therefore a *guillotine type slicer*, originally developed for producing tissue slices for metabolic experiments [M 39], was adopted and modified by LOENING [L 35] to meet the requirements of gel cutting, namely slow but hard pressured slicing. The device chops the gel into slices 0.1–1 mm thick with 0.1 mm intervals, one slice at a time (available from firm 21). Prior to cutting, the gel is frozen in closefitting troughs with powered dry ice (solid CO_2). A similar *macrotome* allowing 15 slices thicknesses between 0.125 and 2 mm was designed by GRESSEL and WOLOWELSKY [G 35].

GROVES et al. [G 46] constructed an *automatic electronic device* for controlled sectioning of cylindrical gels into fractions of 1 mm thickness or less while maintaining them in a rigid state at low temperature. A modified slicer of this type is shown in Fig. 35.

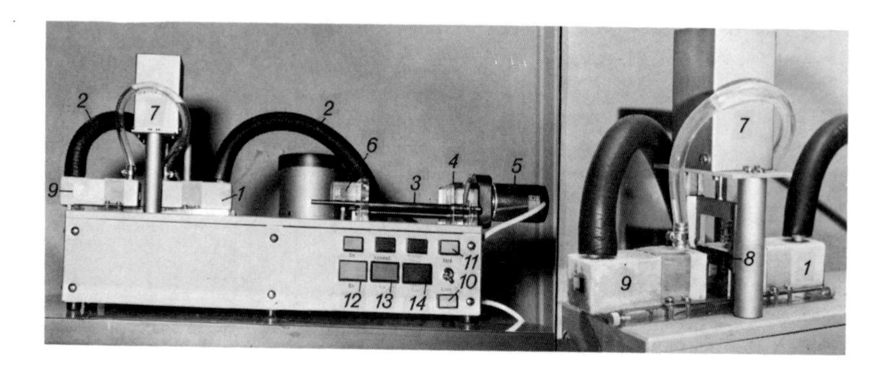

Fig. 35: Automatic gel slicing machine to produce uniform sections of both cylindrical gels and gel strips of any gel concentration and thickness down to 0.25 mm. Acc. to the principle of GROVES et al. [G 46] modified [M 31].

The gel is placed in a double piece metal trough, which is pushed into the jacketed cooling chamber (1) connected with a cooling bath (Kryomat) by insulated tubes (2). The troughs are exchangeable to meet different gel dimensions. After freezing, a removable gel plunger is inserted into the chamber and connected to the drive mechanism which consists of a nonremovable plunger (3), Plexiglas plunger end-plate with quick-release threaded half block (4), synchronous drive motor (5) and a limit switch (6); it advances the gel for the desired slice thickness and stops. This is accomplished by cam gears with appropriate dwell angles. The solenoid (7) is activated and pulls the blade (8) through the gel. The blade is positioned between two separate cooling chambers (1) and (9). When the blade has retreated (by spring action) the drive motor is again activated. This is controlled by a timing motor. The gel sections are collected in the cooling chamber (9). Switches: Automatic (10), Manual (11), On (12), Off (13), Manual sectioning (14).

Ethylenediacrylate cross-linked gels may be used in order to dissolve them in piperidine and Hyamine 10-X according to CHOULES and ZIMM [C 34] (p. 97) for determination of radioactivity. 10 % – gels containing 0.5 % cross-linker are frozen in a cooling jacket by methanol at $-34 \pm 2°$ C and sliced by the regulated movement of a razor blade. Advancing and sectioning are controlled by a single timing motor which does not permit the gel to advance during the sectioning operation. Gels with a wide range of acrylamide concentration can be fractioned.

It should be mentioned that for sectioning the *temperature* of the frozen gel appears to be critical. Too low temperatures yield brittle gels. Ice crystal formation should be minimized. However, no detailed studies were reported so far on optimal temperatures.

A simpler instrument fractionating even dilute (2.7 %) gels into 1 mm thick slices was constructed by BIRNBOIM [B 29]. Freezing prior to fractionation was found to be unnecessary, since the gel is held in a *vertical* position in a cylinder and floated upward to be cut by a moving blade. However uniform slices less than 1 mm thick may be difficult to obtain with this design. A similar, even simpler device was described by IANDOLO [I 1].

MAIZEL [M 3] designed a mechanical gel fractionator on the basis of the following principle: By *extrusion through a narrow orifice* at a constant speed, the gel is crushed and then entrained by a flow of carrier fluid into counting dishes. An electric motor produces a uniform pressure on the gel. Fig. 36 shows the

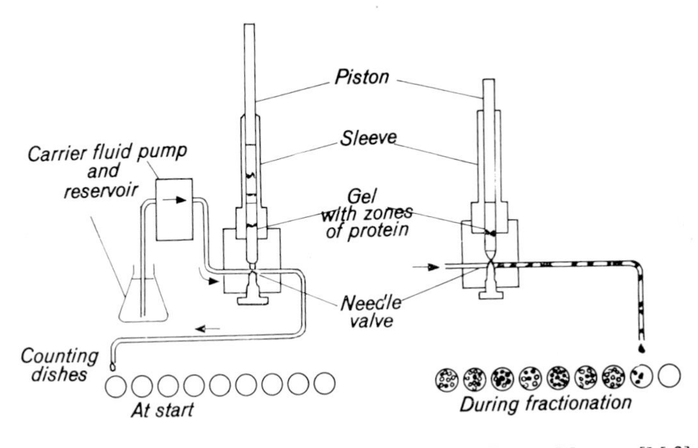

Fig. 36: – Diagram of a mechanical gel fractionator. From: MAIZEL [M 3].

diagram of the fractionator (available from firm 28). Fig. 37 demonstrates a good agreement between three different sectioning methods; in the first two, the constant specific radioactivity of labeled adenovirus proteins serves as a measure for the quantity of substance.

2.3.11. Detection of Radioactive Substances

Since disc electrophoresis represents a microanalytical separation and identification method, special problems arise with regard to the efficiency and detection limit when radioactively labeled substances are to be analyzed on a microscale. The choice of methods of detection described below will depend on the isotopic species (energy, halflife), radioactivity (Curies) and the distribution of the specific activity over the separation distance. Information on radioisotopes in common use in biochemistry and clinical chemistry, their properties, applications and activity measurements can be found, for example, in the books and monographs of BIRKS [B 28], KING and MITCHELL [K 18], SCHWIEGK and TURBA [S 30], SCHÜTTE [S 26], WANG and WILLIES [W 3], WENZEL and SCHULZE [W 13] and WOLF [W 26].

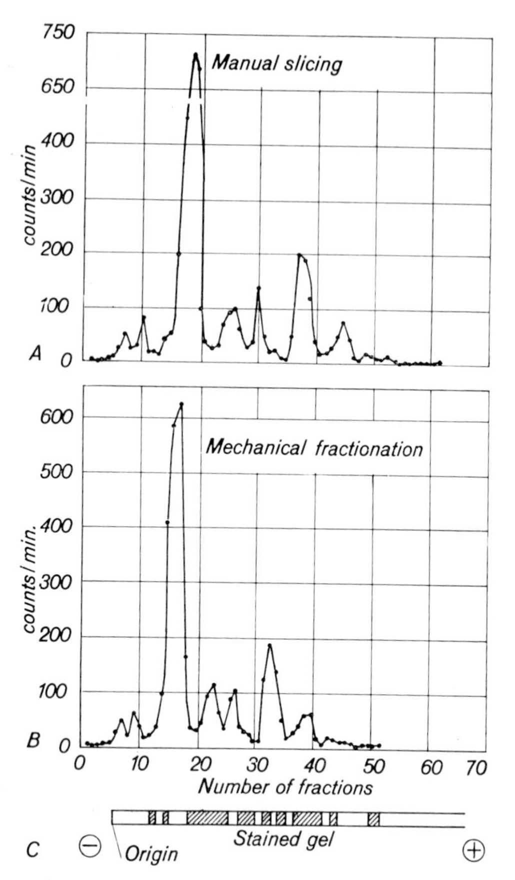

Fig. 37: – Comparison of the patterns of type 2 adenovirus proteins after (A) manual slicing (slicing of the gel and homogenizing), (B) mechanical fractionation with the Maizel fractionator (Fig. 37), and (C) staining with Coomassie Blue (p. 74). From: MAIZEL [M 3].

2.3.11.1. Radioactivity Assays in Gel Sections

2.3.11.1.1. Direct Counting

The simplest method is direct counting of the dried or wet gel sections on dishes, for example, with the use of a 2π-gas-flow counter [F 25]. Self-absorption must be taken into account in this procedure. [M 72]. The method therefore requires relatively high activities to assure for a statistically significant assay. SCHRAM and ROOSEN [S 21] attempted to circumvent the problem of self-absorption by the use of very thin sections.

A more elegant solution was found by BOYD and MITCHELL [B 51] with their method of *water substitution*: The proteins are first fixed with phosphotungstic acid and glacial acetic acid and the gel water is then replaced by scintillation liquid (on toluene basis); for this purpose, the gel water is displaced with a series of organic solvents of decreasing polarity. The loss of labeled protein due to the solvent series amounts to 3 %, while the counting efficiency for ^{14}C in 7.5 % polyacrylamide gel amounts to 81 ± 12 % and for ^{3}H to about 23 %. With the use of a simple "soaking" tank, it is possible to prepare 74 gel sections for counting in 9 to 18 hrs., depending on their size.

2.3.11.1.2. Elution Method

The individual homogenized gel fractions are exhaustively extracted with water or buffer solution. The extracts are centrifuged or filtered. Finally, the radioactivity of an aliquot is assayed directly or after concentration [H 16, Z 3]. However for reasons given below, this method is not generally recommended.

The yields vary highly depending on the type of substance and its elution properties. In some cases, the elution yield can be increased by freezing the gel prior to elution. It should be recalled that large molecules and molecular aggregates diffuse more slowly from the gel than the smaller species; this may lead to errors in a quantitative activity determination if the elution periods are not sufficiently long. For the identification of the fractions, a section of gel is sliced off longitudinally with a razor blade and is stained. It must be kept in mind that the length of stained gel strip may change; this change is taken into account by suitable corrections: The stained gel is divided into an identical number of fractions as the unstained portion and the fractions are compared. An unfavorable factor is that large volumes of eluate are obtained which must be concentrated [L 24]. To avoid differential extraction by diffusion, extraction by electrophoresis is the method of choice.

2.3.11.1.3. Combustion Methods

If the separated substances have low activities or the possibility for detection approaches the lower limit, the following method may be recommended. The stained protein bands (including those from microgels, p. 112) are cut from the gel and heated in Supramax glass capillaries (0.5 mm diameter) with zinc and potassium perchlorate for 45 min. at 650° C according to the combustion method of KOENIG and BRATTGÅRD [K 26]. The tritium activity is counted in the gas phase (isobutane-cyclobutane) with an efficiency of about 80 %. This method is used by HYDÉN et al. [H 49] for tritium counting of very small quantities of protein (10^{-7} to 10^{-9} g) of high specific activity.

An elegant procedure for the determination of ^{3}H, ^{14}C and ^{35}S, however, is the *microcombustion method* of GUPTA [G 47] which uses scintillation vials instead of 1–2 l combustion flasks [K 2, O 5] (Fig. 38). Although the method was designed for biological material (tissue sample of 2–3 mg dry weight and 10–15 mg wet weight), it is also highly suitable for polyacrylamide gels [M 29]. In principle, the gel fractions, dried on lens paper, are combusted in a platinum basket in an oxygen atmosphere. The $^{3}H_2O$ formed is condensed on the inside wall of the vial, while $^{14}CO_2$ and $^{35}SO_2$ are bound by 2-phenylethylamine. The liquids are mixed with scintillation liquid and counted. Reproducibility: ± 2–3 % s.d. Attainable counting efficiences: for ^{3}H about 25 %, ^{14}C about 75 % and ^{35}S about 70 %.

Each gel fraction is dried on a small piece of optical lens paper (1 cm^2, with an attached triangular tip). The paper edges are folded over by pushing the paper into a groove and folding it with tweezers. The triangular tip is blackened with ink. The sample is then placed into a platinum basket consisting of wide-mesh Pt wire screening and resting on a Pt wire coil; the unit is transferred into a vial and flushed with oxygen. The cover is closed tightly and the inked tip is ignited from outside with a projector lamp (e.g. 8 V, 50 W, mounted on an electrical soldering iron). The sample must burn completely without an ash residue. It is important to permit all of the oxygen to diffuse to the sample without obstruction (this is the reason for the wide-mesh screen which is located exactly in the center of the vial). By dipping the vial into liquid nitrogen or a dry ice-acetone bath the formed water is condensed. Finally, the vial is opened and filled with scintillation liquid (e.g. 7.0 g PPO, 0.3 g dimethyl-POPOP, 100.0 g naphthalene, and dioxane to 1 l. Supplier of scintillation chemicals are, for example, the firms 1, 20, 24, 30). A fiber-glass disk soaked with 2-phenylethylamine is placed on the bottom of the vial prior to flushing with O_2 for the absorption of $^{14}CO_2$ and $^{35}SO_2$.

McEven [M 36] described a similar technique using a V-shaped Pt-stand. To determine specific activities, protein content is assayed.

The stained and scanned gel is scored where it is to be cut, then again scanned with a densitometer equipped with an integraph. The integrated optical density of each gel segment is converted to protein amount by means of a calibration curve relating integraph units to protein amount. Finally the gel is cut and combusted to give tritiated water. Counting efficiency: around 35 % for ^3H.

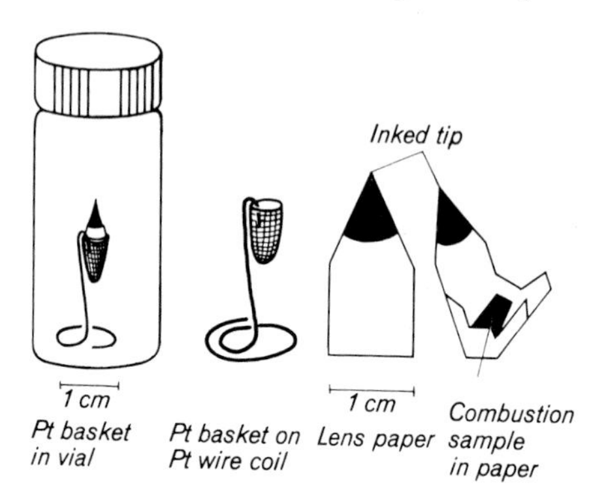

Fig. 38: – Equipment for microcombustion of gel fractions according to Gupta [G 47]; modified [M 29].

2.3.11.1.4. Hydrolysis Method

The following methods share the characteristic that the cross-linkage of the gel is broken by hydrolysis leading to a collapse of the gel structure. In the first method of Shepherd [S 39], hydrolysis is carried out with *concentrated hydrochloric acid*. The proteins are cleaved into peptides and amino acids; the dyes usually no longer exhibit a quench effect.

After staining, densitometric analysis and slicing of the gel, each gel segment is hydrolyzed in a fused ampule (Duran, Pyrex) in 1 ml 6 N HCl for 16 hrs. at 120° C (autoclave). After neutralizing with 10 N NaOH and cooling, the material is centrifuged (2000 rpm, 2 min.). Drops of the supernatants are placed on fiberglass disks in glass vials for liquid scintillation counting. After drying of the solutions, toluene-containing scintillation liquid is added.

A more elegant and simpler method was developed by CHOULES and ZIMM [C 34] which, however, can only be applied for neutral and acid gel systems, e.g. for the study of basic proteins. If N,N'-methylene-bis-acrylamide, the cross-linking agent, is replaced by *ethylenediacrylate*, which contains ester groups instead of the amide bonds, a complete hydrolysis can be obtained under the influence of aqueous piperidine, leading to solubilization of the polyacrylamide gel in organic solvents. See also page 60 for other cross-linkers yielding solubilizable gels.

Immediately after electrophoresis, the gel is frozen in dry ice, cut, and the fractions are transferred into 20 ml counting vials. After addition of 0.5 ml of a mixture of 9 vols. 1 M piperidine and 1 vol. Hyamine 10-X, the vial is shaken for 1–4 hrs. (depending on the quantity of gel) at 37° C until dissolution. Finally, 10 ml scintillation liquid are added. The quantity of water approaches a critical limit in this case. The vials must be tightly closed to prevent any volatilization. If difficulties arise, more (0.1 ml) 1 M piperidine is added and Hyamine 10-X only after hydrolysis (rather than before). Scintillation liquid, for example, according to KINARD [K 17]: 576 ml xylene, 576 ml dioxane, 348 ml absolute ethanol, 120 g naphthalene, 7.5 g PPO and 75 mg dimethyl-POPOP. According to Choules and Zimm [C 34], the counting efficiency for ^{3}H amounts to about 10 % and about 98 % of the radioactivity is recovered. Suppliers for Hyamine 10-X are, for example, the firms 24, 30.

Instead of using piperidine and Hyamine, SPEAR and ROIZMAN [S 62] hydrolyze each gel segment in concentrated ammonia and adsorb the sample on a glass fiber disk which is dried and treated with scintillation liquid. Similarly using ammonia, CAIN and PITNEY [C 2] determine the relative specific radioactivity of ^{3}H-proteins in each gel section after disc electrophoresis at pH 8.9 in an ethylenediacrylate cross-linked gel.

The gel is stained, segmented, each fraction dissolved in 1 M NH_4OH (400 mm^3 of gel / ml ammonia) for a few hrs. and the protein-bound dye measured colorimetrically (at 610 nm for Amido Black stained gels). The ammonia is evaporated to dryness at 65° C. Then 0.2 ml of water are added to rehydrate the gel which is dissolved in 1.5 ml of NCS (a solubilizer available from firm 1) for few hrs. at room temperature. 10 ml of scintillation liquid (4 g PPO + 0.3 g Dimethyl-POPOP/ l toluene) are added. Counting efficiency for ^{3}H: 26–30%. – Moderately firm gels crosslinked with ethylenediacrylate at slightly alkaline pH can be obtained with tris-HCl buffers, but not with tris-EDTA-borate buffers.

ALPERS and GLICKMAN [A 24b] prefer NaOH to solubilize ethylenediacrylate cross-linked gels since by means of NaOH protein quantities from 10 to 60 µg may be accurately determined according to LOWRY et al. This procedure overcomes the dye-binding problem (see page 82) but introduces the drawbacks of the Lowry method.

Gel segments containing total protein bands are cut and fixed in 5% TCA for 8–12 hrs. to remove all residual Tris, then rinsed with water, blotted dry and dissolved in 0.3–0.4 ml NaOH. All samples are brought to equal volumes (0.5 ml). 0.1 ml aliquots are used for protein determination according to LOWRY et al. [L 41a]. 0.2 ml aliquots are added to 1 ml NCS and heated at 65° C for 1 hr. in scintillation vials. Finally 10 ml scintillation liquid are added.

A similar method was used by LEWIS et al. [L 29a] to estimate the protein quantity in the growth hormone and prolactin bands of disc electrophoretic patterns. These authors observed that the stabilizer contained in any ester batch changes the Amido Black stain and should therefore be removed by stirring the liquid of a newly opened bottle for 2 hrs. at 20° C in a well-ventilated fume hood. Then the material was found to remain stable if kept in the dark at 5° C. Moreover, the working solutions should be prepared immediately before use and electrophoresis performed at 5° C.

However these gels are not completely stable at the alkaline pH of the buffer due to hydrolysis of the ester linkage. In addition they are reported [C 2] to be less firm than Bis-crosslinked gels and thus more difficult to cut uniformly. They provide satisfactory resolutions, although the patterns are not quite as sharp as those obtained in Bis-gels.

2.3.11.1.5. Free Radical Depolymerization Method

A rapid collapse of the gel structure can also be obtained by depolymerization with 30 % *hydrogen peroxide* according to YOUNG and FULHORST [Y 5]. This method yields satisfactory recoveries for ^{35}S-containing proteins [Y 5]. To avoid the risk of activity loss due to the suspected formation of $^{14}CO_2$ and $^{3}H_2O$ MOSS and INGRAM [M 70] use rubber caps through which they inject Hyamine 10–X following hydrolysis in order to bind carbon dioxide and water.

However TISHLER and EPSTEIN [T 15] found no conversion of ^{3}H to $^{3}H_2O$ when they solubilized 1.27 mm thick gel slices by incubating the well dried slices in 0.1 ml of 30 % H_2O_2 at 50° C for 1–6 hrs. These conditions also yield less than 5 % loss of ^{14}C due to formation of volatile $^{14}CO_2$. The capacity of this method is demonstrated by Fig. 39.

After completion of solubilization 1.0 ml of NCS (see above) and 10 ml of scintillation liquid (3.0 g PPO + 0.1 g POPOP/l toluene) are added. Counting efficiencies: for ^{14}C 80 %, for ^{3}H about 23 %. Recoveries of isotopes usually exceed 90 %, but fall below 90 % in slices of highly concentrated gels (15 % acrylamide).

Similarly LE BOUTON [L 11] observed little loss of label due to peroxide treatment if the peroxide-liberated, labeled protein was solubilized by NCS.

GRAY and STEFFENSEN [G 32] solubilized their paraffin embedded gel slices (p. 91) in 30 % H_2O_2 and count them in BRAY's solution [B 55]. The counting efficiency for ^{14}C and ^{35}S is about 90 % and for ^{3}H about 33 %. Their method is also applicable for analysis of double-labelled compounds in gels.

Each gel is treated with about 5 ml of toluene to solubilize the paraffin surrounding each slice. After removal of the toluene 10 min. later, 0.1 ml of 30 % H_2O_2 is added to each vial, which is tightly kept to prevent evaporation and tilted to insure good contact with the small volume of H_2O_2. When each slice is completely solubilized after about 3 hrs. at 60° C or overnight at 20° C, 20 ml of BRAY's solution

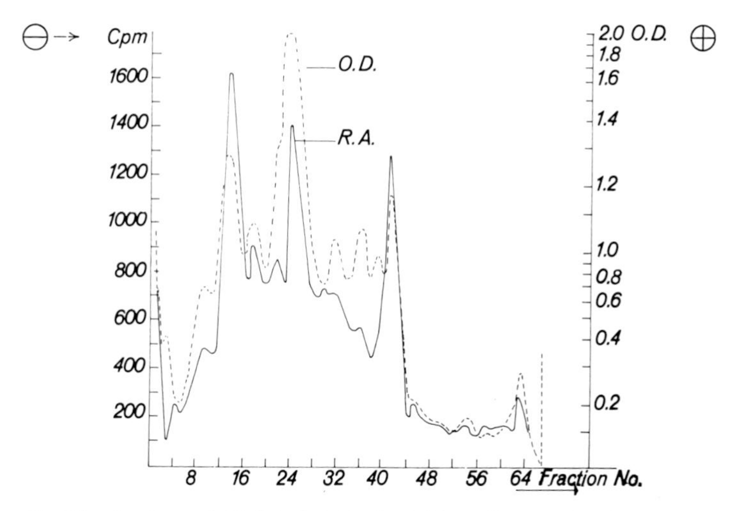

Fig. 39. Distribution of ^{3}H-labeled Escherichia coli proteins in a disc electropherogram.

E. coli strain CA–34 were grown in the presence of ^{3}H-leucine, proteins isolated acc. to NEU and HEPPEL [N 26] and electrophoresed in gel system No. 1a at 3,5 mA/gel (5 mm ø) for 55 min. Amido Black staining. Electrophoretic destaining. Scanning of the gel using the microdensitometer from firm 17 with filter (O. D.). Sectioning: 0.75 mm slices were cut with the automatic gel slicer of Fig. 35 at –18° C, dried and dissolved in 70 μl of 30% H_2O_2 at 55° C for 5 hrs. in closed vials. Then 150 μl NCS (see page 97) and 3 ml toluene-based scintillation liquid were added and the radioactivity (R. A.) determined.

[B 55] are added and the radioactivity determined. This procedure was found to be superior to desolving in Hyamine 10-X dried residues of slices solubilized in H_2O_2 and counting in PPO-POPOP-toluene scintillation liquid, since the net counts of the slices decrease by 40–50 % after 24 hrs.

BASCH [B 10] found it unnecessary to break down the gel structure in order to solubilize fixed labeled material. When Bis-cross-linked gel slices are treated with NCS, the sections rapidly swell and release entrapped macromolecules into the liquid phase under hydrolysis. Counting efficiencies attainable for ^{3}H up to 40 %, for ^{14}C up to 82 %.

1–2 mm slices are completely covered with 0.5 ml NCS, incubated at 65° C for 2 hrs. in test tubes without agitation and decanted along with NCS into scintillation vials. Following rinsing with 3 ml of toluene based scintillation liquid additional 12 ml are added. For complete elution the vials are allowed to stand at 0° C overnight before counting. Recovery: > 90%. Highly concentrated (15%) gel slices resist swelling by this procedure but may do so after addition of 10% water to NCS [Z 1a].

2.3.11.2. Autoradiography of Longitudinal Gel Sections

Developed by FAIRBANKS et al. [F 1], this technique can be recommended especially for its high detection sensitivity of proteins of high specific activity; protein quantities that can no longer be localized with the usual staining methods may still be identified.

After staining and destaining, the gel is cut into 4 longitudinal strips with the use of the slicing device shown in Fig. 40; the two inner slices have a thickness of 1.6 mm each and should have a uniform, plane surface. These two inner gel strips are spread on wetted filter paper which is placed on a small-pore suction filter. The top side of the gels is covered completely with a thin flexible plastic film stretched smoothly over the gel strips. The side edges are sealed with a flexible rubber gasket and silicone grease to permit application of a vacuum. This arrangement mechanically fixes the gels and prevents them from shrinking. They are evacuated for 1–2 hrs. with irradiation from an infrared lamp (at a distance of about 30 cm). The thin, hard and dried strips are now covered with Kodak No-Screen medical x-ray film (for ^{14}C), clamped firmly between two low-background metal plates and stored light-protected in a cool place. Standard reagents for processing Kodak x-ray film are used. During developing it is advisable to cover the unexposed side of the film to prevent an increase of background radiation; of course, the plastic film must be removed during fixing. Equipment for this technique is available from firm 5.

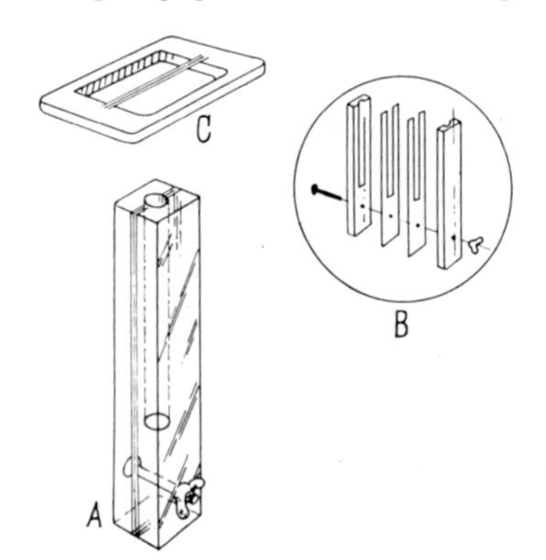

Fig. 40: Apparatus for longitudinal slicing of cylindrical polyacrylamide gels. From: FAIRBANKS et al. [F 1].

(A) Plexiglas holder (2.5 × 2.5 × 15.2 cm, length of bore 10 cm, bore diameter 7 mm). (B) Holder components. (C) Cutting tool of brass with 3 taught stainless steel wires of 1.6 mm spacing.

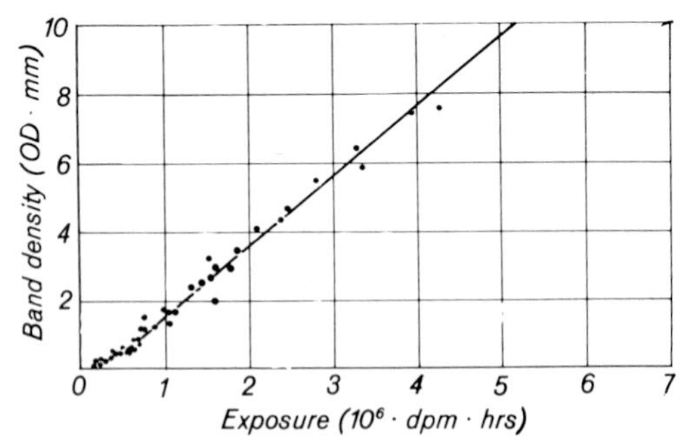

Fig. 41: Calibration curve for the quantitative autoradiographic assay of ^{14}C-labeled hemoglobin in disc electropherograms. From: FAIRBANKS et al. [F 1].

Exposure expressed in duration [hrs.] x (estimated number of dpm in *whole* gel); the linear band density ([mm] × optical density unit) corresponds to the area under the microdensitometer curve of the autoradiograms.

According to Fig. 41, a linear relation exists between the number of radioactive disintegrations of a protein subjected to electrophoresis and the degree of blackening expressed as the band density of the photometrically analyzed autoradiogram. – Since the dried gel strips theoretically have an infinite thickness for ^{14}C and ^{3}H under the described conditions, fluctuations in gel thickness have less an effect than fluctuations of the water content which can cause considerable variations in counting efficiency. It is therefore advisable to maintain constant drying conditions. – Fig. 42 illustrates the advantage of autoradiography for the detection of microquantities of protein of high specific activity compared to the staining method (compare B 2 with C 2). Even if only low specific activities per protein band are present, detection may still be successful by using longer exposure times.

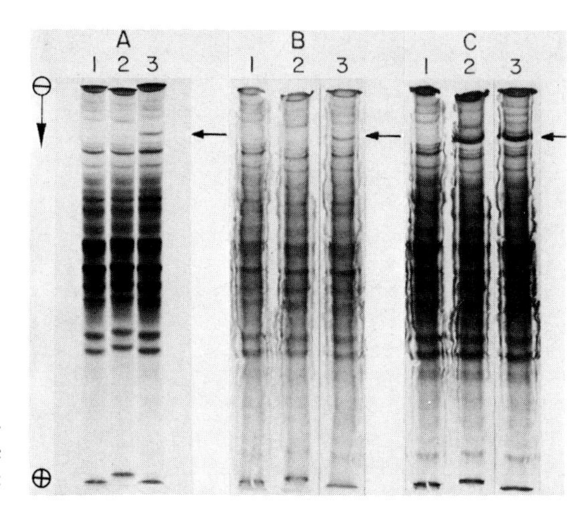

Fig. 42: – Disc electrophoresis of Tris-soluble proteins from *E. coli*. From: FAIRBANKS et al. [F 1].

(A) Unsliced stained gels; (B) dried strips of stained gels; (C) autoradiograms of B (exposure 81 hrs.). Three cultures each were labeled with 20 $\mu C\,^{14}C$-leucine (160 mC/mmole) for a 4 min. pulse, after which 80 % were incorporated in acid-insoluble materia . In each gel, $5 \cdot 10^5$ dpm of the material were subjected to electrophoresis. – Culture 1: Control (not induced). Culture 2: induction of β-galactosidase by addition of isopropyl-β-D-thiogalactoside ($5 \cdot 10^{-4}$ M culture concentration) 2 min. prior to labeling. Culture 3: induction as in culture 2 but 25 min. prior to labeling. – 7.5 % polyacrylamide gels, Amido Black stain. Length of separation gel 98 mm. The arrow marks the location of the β-galactosidase (identified with o-nitrophenylgalactoside).

JOHN and MILLER [J 7] detected ^{35}S-labeled serum albumin by a similar method. The authors dispensed with drying of the gel strip. Instead, the gel halves, obtained by a simple longitudinal cut, were wrapped air-tight in thin plastic films. Kodak Blue medical x-ray film, placed on the plastic-covered inner sides of the gel halves, i. e. the smooth surfaces of the semi-cylinders, was exposed for 6 weeks at 4° C.

2.3.12. Drying of Polyacrylamide Gels for Storage or Radioactivity Determination

Besides the method by FAIRBANKS et al. [F 1] as described above, the following procedures may be used in order to dry polyacrylamide gels for storage or radioactivity determination.

According to HERRICK and LAWRENCE [H 20] gel strips are immersed in an aqueous solution of a nonionic detergent (e.g. Tween 80, 0.02 %) and glycerol

(0.3 %) for 15 min., rinsed briefly with water to remove the excess of detergent and spread on a smooth piece of Plexiglas. The gels and Plexiglas are introduced into a dialysis tube which is stretched over the gel allowing no bubbles to form. Drying proceeds for 12 hrs. at 70° C and the gel is removed from the Plexiglas support by placing the unit on a flat plate, gel side turned down, and cutting the dialysis tube open on the side. The exposed gel side finally can be covered with an autoradiography film. The dried gels are wrapped in Cellophane for storage.

Lim et al. [L 31] place the gel on a wet dialysis membrane which, in turn, they overlay with porous plastic disks fitted to the base of a bell jar. Gel, dialysis membrane and plastic disks are fixed to the jar with Saran Wrap tightened by rubber bands. Drying takes place under evacuation and warming with an IR-lamp.

According to Gabl and Pastner [G 1], *permanent specimens* with a stability for several years can be prepared as follows (see. also [S 52]).

The gel strip is placed on a wet Cellophane film, mounted on a firm and smooth porous support (unglazed clay slab). A second wet film is then placed over the strip, the ends are tightly folded over the edges of the support and fastened. Drying takes place at room temperature for 12–14 hrs.

Another drying procedure was reported by Reid and Bieleski [R 21]. Polyacrylamide gel slabs are placed in a tray on a clean plate of glass sufficiently large to allow a 2 cm margin all around the gel. 2 % molten agar is poured over the gel and glass plate in a uniform 4 mm layer and allowed to set. The glass plate is placed under an IR-lamp and dried slowly and evenly until almost dry, then put aside for final drying at room temperature.

Borris and Aronson [B 48] pointed out that the gels can be uniformly shrunk and later fully rehydrated without distortion of stained patterns. This may be achieved by drying e.g. at 55° C and 300 mm Hg in a vacuum oven for 4–5 hrs. The gels should be prevented from touching any surface to which they might adhere during the drying process. Therefore the gels are suspended from the oven rack by a stainless steel wire, inserted through the lower part of the gel below the tracking dyeband. Fading of band due to dye diffusion into storage solvent does not occur when the gels have been dried. The gels easily reswell in water or 7 % acetic acid at 70° C in 3–4 hrs.

Daniels and Wild [D 2b] found that one side of a gel slab can be firmly bonded to a rigid surface permitting subsequent drying without distortion or shrinkage. The bonding agent Aerolite 306 synthetic resin glue (Ciba Ltd.) is mixed with water and spread over the roughened surface of a hard wood piece. The gel slab is immersed for a few min. in 30% (w/v) aqueous formic acid, placed on the resin-coated wood and left for 2–3 days at room temperature to dry.

2.4. Disc Enzyme Electrophoresis

Since the current generates more or less heat during electrophoresis (page 21), depending on the voltage gradient used, thermolabile enzymes are subjected to electrophoresis at low temperatures and low current intensities (e.g. 4° C, 1–3 mA/gel); this, however, requires longer running times (e.g.

100–150 min.). – In principle the following *enzyme detection techniques* can be distinguished.

2.4.1. Elution Method

This simple method consists of cutting the gel, homogenizing the fractions with suitable buffer solutions, and testing the eluates for enzyme activity. This method corresponds to that described on page 95, to which the reader is referred for pertinent comments.

2.4.2. Gel Staining Method

In this case, the enzyme-containing bands are identified by incubating the gel in substrate solution, which gives rise to an enzyme-specific color reaction. The method requires freely diffusible substrate molecules. In practice, the gel is rinsed with distilled water immediately after electrophoresis and incubated in about 3 ml substrate solution. Microdensitometers with filters for the particular dye-complexes involved are suitable for quantitation. For a quantitative analysis, the recorded absorbance values must be referred to enzyme activities; calibration curves are therefore constructed with defined quantities of enzymes or activities. The stain development must be reproducible, however. The mixing procedure of the reagent solutions, incubation temperature and time should therefore be standardized.

This method is also applicable for the detection of transferases. Thus, to assay RNA *polymerase*, Krackow et al. [K 34] incubate the electrophoresed gels with substrate (ATP + UPT) for unprimed polyA-polyU synthesis, wash and stain with ethidiumbromide which is specifically bound by the enzyme product.

To locate *enzymes converting nonreducing substrates into reducing products* such as sugars in disc gels, Gabriel and Wang [G 2] modified the formazan reaction and developed a semiquantitative evaluation of enzymic activity even of crude extracts. The enzymes detectable by the technique include: hydrolases (e.g. sucrose phosphorylase), glycosidases (e.g. glucosaminidases) and oxidoreductases (e.g. TDP-glucose oxidoreductase). In addition, any macromolecules (e.g. glycoproteins) susceptible to periodate oxidation can selectively be stained thereby. Finally the technique permits an easy correlation of proteins with enzymes in the same gel, since the formazan band retains its appearance in the presence of superimposed protein stain.

2.4.3. Substrate and Primer Inclusion Techniques

The gel staining method described above is based on rapidly diffusing, low molecular weight substrates – a prerequisite which cannot be satisfied for a number of substrates (e.g. DNA, RNA, starch). In such cases, the substrate is polymerized into the separation gel, the enzyme solution is subjected to electrophoresis at temperatures of about 0° C and incubation proceeds with substrate specific stains under optimal conditions for enzyme activity [B 50, D 23–24, L 30a, S 41a, S 66a]. Naturally, any enzyme activity must be suppressed during electrophoresis. Fig. 43 presents a schematic diagram of the principle of the *substrate inclusion technique* using the detection of deoxyribonuclease as an example.

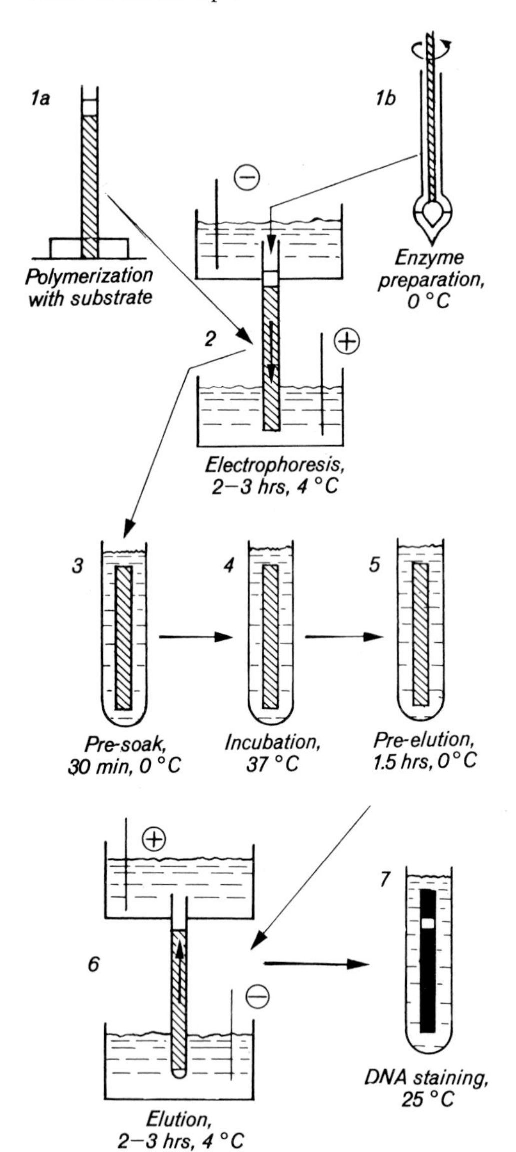

Fig. 43: – Schematic diagram of the reaction steps for the detection of deoxyribonuclease. Enzyme electrophoresis in DNA-containing polyacrylamide gel. From: Boyd and Mitchell [B 50].

Preincubation (step 3) after enzyme electrophoresis (step 2) leads to optimal buffer conditions in the gel. Pre-elution (step 5) and electrophoretic elution (step 6) remove reaction products (oligonucleotides) and a part of enzyme proteins (nucleases).

While the substrate inclusion method is mainly used to identify hydrolyzing enzymes, the *primer inclusion technique* permits the detection of synthesizing enzymes which are primer dependent. As the former method, the primer inclusion technique utilizes primer molecules incorporated into the gel in an immobilized state which is maintained during electrophoresis of the enzymes. The technique is highly sensitive since down to 10^{-12}g enzymes may be detected in microgels (see page 114). Examples of both methods are listed in Table 12. For micro-enzyme analysis after micro-disc electrophoresis see page 114.

Table 12: Analysis of enzymes after polyacrylamide gel electrophoresis: Examples for the substrate and primer inclusion technique. From: STEGEMANN [S 66b].

Enzyme	Gel-included substrate or primer	Precursor	Dye to stain gel-included substrate or primer	Reference
A. Substrate inclusion technique				
Amylase	starch	—	iodine	S 41 a–b
Deoxyribonuclease	DNA	—	Methylgreen	B 50
Hyaluronidase	chondroitin-4-sulfate	—	Alcian- and Toluidine blue	L 30a
Pectinase	pectic acid	—	Methylene blue, ruthenium red	S 66a
Phosphorylase	starch	—	iodine	S 41b
Proteinase	collagen	—	Amido black 10 B	
B. Primer inclusion technique				
Phosphorylase	glycogen	G-1-P	iodine	S 41 a–b
Polynucleotide-phosphorylase	poly A poly U s-RNA	ADP UDP nucleoside-diphosphate	Methylene blue Pyronine	S 66 b
DNA-Polymerase	poly d(A-T)	d ATP d TTP	Pyronine	N 28
RNA-Polymerase	poly d(A-T)	ATP, UTP	Pyronine	N 31–34

2.4.4. Contact Print Method

This mild procedure is recommended for enzymes which, as in the preceding case (chapter 2.4.3), are to convert high molecular weight substrates but should be directly detectable in the gel without modification [D 25–26]: they can then be further analyzed with other techniques.

After electrophoresis, the gels are placed into the grooves of an undulated plastic mold and covered with a glass plate which is coated with a substrate-polyacrylamide film (for example, for amylases, separation gel with 1.5 % hydrolyzed Connaught starch, pH 7.4). The plate is placed under a weight, is incubated and developed with a reagent (e.g. iodine solution) which stains unconverted substrate (Fig. 44). The enzyme bands and even the gel contours contrast readily on the plate; the bands can be quantitated by densitometry. With proper packaging, the developed plates are stable without modification for several months at 4° C. – Finally, the enzyme-containing bands can be cut from the gels and eluted.

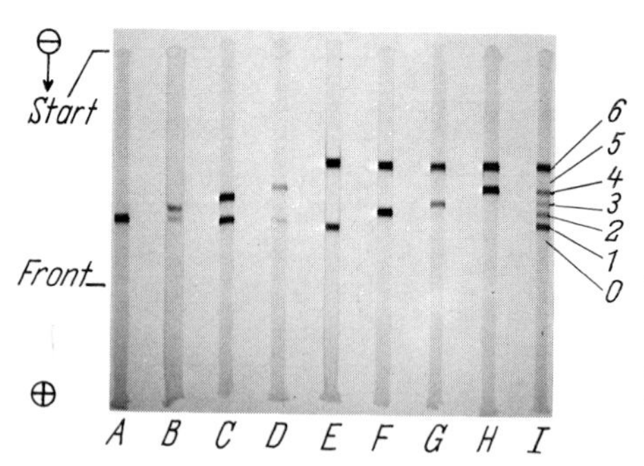

Fig. 44: – "Contact print" of a stained starch-polyacrylamide film: Amylase-isozyme pattern of different *amy* strains of female *Drosophila melanogaster* after disc electrophoresis. From: Doane [D 26].

(A–H) Individual homozygotes; (I) mixture of all alleles (each in 1/8 concentration). Bands 0 and 5 are barely visible but can be recorded by densitometry.

2.5. Disc Immunoelectrophoresis

2.5.1. Two-dimensional Disc Immunoelectrophoresis

In this procedure, the sample material is subjected first to disc electrophoresis (first dimension) and then to immunodiffusion (second dimension): The antibodies thus are allowed to diffuse perpendicular to the antigens which were separated in the running gel (Fig. 67). Both agar and polyacrylamide gels are suitable media for immunodiffusion.

2.5.1.1. Immunodiffusion in Agar Gel

Three agar immunodiffusion techniques performed after gel electrophoresis may be distinguished according to Darcy [D 3]. See also [B 14].

2.5.1.1.1. Plain Agar Method

The electrophoresed gel strip, obtained by slicing the cylindrical gel lengthwise (p. 100), is laid on a 1–2 mm thick agar layer. Antibody troughs 2 mm wide are cut immediately along either side of the strip using a scalpel to which a second blade has been attached. Precipitate lines form in the agar underneath the strip. A reaction of identity (or no identity) between bands of two different strips may be obtained by laying the strips on agar side by side, cutting troughs along the outer edge of each strip and allowing immunoreaction from each side. Resulting precipitate lines underneath the strips will either fuse (identity) or cross-over (non-identity).

2.5.1.1.2. Antibody-Agar Method

The gel strip is put on a bed of antibody containing agar 1–2 mm thick. The antiserum and agar are mixed (1 : 4 e.g. for the study of albumin) and poured at 45° C. As early as 15 min. after the strip is laid on the agar the spots of precipitate are visible. The presence of distinct bands is shown less ambiguously by the spots than by the usual precipitation lines. A disadvantage may be that high antibody concentrations are required.

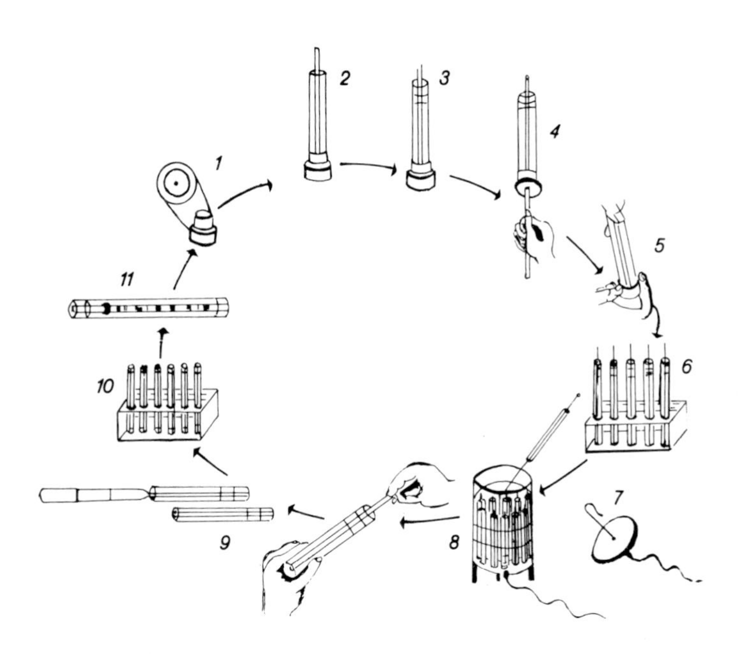

Fig. 45. Stepwise procedure of disc immunoelectrophoresis using hollow gel cylinders acc. to Makonkawkeyoon and Haque [M 3a].

(1) Make a hole in the center of a vacutainer rubber stopper. (2) Support empty glass tube and plexiglas rod on the stopper. (3) Fill tube with separation gel. Layer 1 ml of distilled water. (4) Allow to polymerize with another rod. Gently push the plexiglas rod out of the stopper only. (5) Remove rubber stopper and place tube vertically in a rack. (6) Place top gel and riboflavin. Allow to gel by photopolymerization. (7) Place tube in apparatus. Apply sample and electrophorese. (8) Remove rod; place tube horizontally. (9) Fill the lumen with antiserum-agar mixture. (10) Seal both ends with parafilm; keep in refrigerator. (11) Read results.

2.5.1.1.3. Agar Embedding Method

The gel is put on an empty Petri dish, warm agar solution (1.5 % in 0.15 M NaCl at 45–50° C) is poured around it and allowed to gel. Antibody troughs are cut 3 mm from the gel or, if the gel has been covered with the agar to the depth of about 1 mm, the troughs may be cut touching the gel on either side. In the latter case the precipitation lines form in the agar on top of the gel, provided that the troughs have been filled completely. Due to the good contact between agar and polyacrylamide gel sharp and regular lines are observed.

Even simpler, the electrophoresed gel, strip or cylinder, may be gently squeezed into a 3–4 mm wide trough cut in a 2 mm thick agar layer. Antibody troughs 2 mm wide are cut along either side of the gel trough.

An elegant modification was reported by MAKONKAWKEYOON and HAQUE [M 3a]. The polyacrylamide gel is polymerized as a *hollow cylinder* by placing a Plexiglas rod in the middle of the tube (Fig. 45). Following electrophoresis the rod is removed and the lumen is filled with the antiserum mixed with molten agar. The procedure yields clearly separated precipitin bands in the agar gel which are easy to count since they appear as discs rather than as overlapping arcs. Another advantage is that removal of the gel from the tube is no more necessary.

For comparison and identification of protein antigens, one half of the lengthwise sliced gel may be stained with Amido Black, while the other is subjected to immunodiffusion. Alternatively, the polyacrylamide gel is sliced into discs and these fractions are allowed to react with antiserum (Fig. 79). Immunodiffusion usually takes place for several days, depending on the gel size, at 4–10° C.

General information and further details on immunoelectrophoretic techniques can be found, for example, in the books by GRABAR and BURTIN [G 30] or CLAUSEN [C 42a]. BEDNAŘÍK [B 15] described a method for the quantitative evaluation of disc immunoelectrophoretic fractions. ZWISLER [Z 9] and BIEL [B 25] combined horizontal gel slab electrophoresis with vertical immunodiffusion (agar cast onto the polyacrylamide gel). For an agar double diffusion method see e. g. [M 68], for micro-disc electrophoresis combined with micro immunoanalysis see page 115. Supplier of agar for immuno-methods: for example, firm 23.

FELGENHAUER [F 13] pointed out that the normal immunodiffusion reaction necessitates an application of protein amounts 2–3 times greater than that required for ordinary electrophoresis. The amount of antiserum necessary for optimal precipitates (40–80 μl) seems to be inadequate for a micromethod. Since only a minor part of the antibodies is needed for the immunoreaction, Felgenhauer developed the following technique. Subsequent to micro-disc electrophoresis the gels are shaken at 4° C for 17 hrs. in small vessels containing 50–60 μl antiserum. After washing in saline (+ 0.01 % merthiolate) over 4 days at 20° C the gels are stained with Amido Black and destained. The optimal antigen-antibody ratio has to be found by trial and error. The antiserum can be used at least 16 times. This method may be useful, if only small amounts of valuable serum are available.

2.5.1.2. Immunodiffusion in Polyacrylamide Gel

Little difference was found between agar and polyacrylamide gels with regard to the immune reaction [A 29b, K 12]. Nevertheless, polyacrylamide gel offers certain advantages: It is simpler to handle, shows no electroosmosis, has a higher elasticity (agar gel has a tendency to break at the starting point where the sample is applied) and shorter running times are sufficient (about 30 min); a disadvantage is that the immune reaction requires more time (72 hrs. compared to 24 hrs. in agar gel).

A layer of 2.8 % polyacrylamide of 1 mm thickness is used. Parallel troughs of 50 mm length (1 mm width) with a spacing of 12.5 mm are formed in this layer; the distance between trough and starting point on the cathode side is 6.2 mm. – According to ATHERTON and HERSHBERGER [A 44], crystal-clear gels can also be obtained in Petri dishes if the concentration of TEMED is increased from 0.23 ml to 0.66 ml and of ammonium persulfate from 0.14 g to 0.26 g/100 ml solution. This counteracts the inhibiting action of atmospheric oxygen on the polymerization (p. 3).

MENGOLI and WATNE [M 43] prefer polyacrylamide gel to agar or agarose gel as a medium for immunodiffusion, because the synthetic polymer is characterized by chemical inertia, complete transparency and variability of pore sizes. For example, histones show a nonspecific reaction with agar and agarose but not with polyacrylamide.

Polyacrylamide gel does not interfere with antibody induction in the rabbit [S 84]. Thus, segments containing antigens may be pooled, homogenized in 0.5 ml saline and, after addition of 0.5 ml Freund's adjuvant, inoculated s.c. on the back of a rabbit [H 11].

2.5.2. Unidimensional Disc Immunoelectrophoresis

The interactions (precipitation reactions) of antigens and antibodies can be determined qualitatively and quantitatively in a single electrophoresis step with the use of this method. For this purpose, the antibodies are polymerized into the spacer gel and the antigens into the sample gel. Under current, the antigens migrate into the spacer gel where they precipitate by reacting with the antibodies. Non-specific and excess antigens or antibodies enter the separation gel; in this manner, they can be separated from the specific and bound substances.

FITSCHEN [F 24] used this method in the isotopic dilution analysis of ^{131}I-growth hormone with a detection limit of 0.4 ng. The precipitation reaction was expressed in per cent of immunoglobulin bound radioactivity. The bound radioactivity was found to be inversely related to the logarithm of the unlabeled antigen concentration in the sample gel. The activity was assayed by direct counting of the gel fractions (p. 94).

LOUIS-FERDINAND and BLATT [L 40] used direct densitometry to quantitative the degree of interaction between human albumin and purified horse immunoglobulin.

The reaction was stoichiometric up to about 15 μg albumin (gel system No. 1a, sample gel half as long as spacer gel, electrophoresis at 2.5 mA/gel and 21° C). The fraction of unbound albumin which had migrated into the separation gel was measured.

The method is suited in instances when the antigen-antibody reaction can only be determined indirectly, e.g. for haptens; it is also applicable to determine the specifity of different antisera (i.g. partially purified globulin fractions) and to identify and quantitative certain fractions during immunochemical purification procedures. Compared to other quantitative diffusion methods, this procedure has the advantage of a considerable time saving. See also [C 11].

3. MICRO-DISC ELECTROPHORESIS

3.1. General Methods

Separations on a micro- and ultramicroscale by disc electrophoresis of *protein quantities between* 10^{-6} and 10^{-12} *g* have attracted notable interest, particularly in cytology and microbiology. The greater technical expense (micropipettes, micromanipulators, microscopes) is balanced by the following advantages [G 40]: Higher detection sensitivity because of smaller gel cross sections; application of higher voltage gradients because of improved heat dissipation; shorter running times and diminished diffusion due to more rapid fixation and staining of separated sample components.

Pun and Lombrozo [P 32] designed a micro-electrophoresis apparatus for the separation of cytoplasmic proteins from 0.8–2 mg rat brain tissue.

The gel molds consist of vertical glass channels (microcells of 1.8 × 1.0 × 70 mm) cut into a Pyrex glass plate. Thus, the cross section of the gel strips is smaller than $^{1}/_{10}$ of the standard gel cylinder (5 mm diameter). This eliminates the need to collect small quantities of identical tissue from different individuals to obtain the quantities of sample required for electrophoresis. Another advantage results from the concentrating effect (page 22) which permits the use of relatively large volumes of extract. Since the moving boundaris in the gels migrate at different velocities, it was found neccessary to supply each microcell with its own constant regulated current.

Grossbach [G 40] reported on the separation and quantitation of proteins on a *nanogram scale* (10^{-9}g). The micromethod was developed for the analysis of the secretion of individual salivary gland lobes of *Chironomus* larvae [G 42] and was tested with the quantitative detection of 3.3 × 10^{-9} g serum-albumin dimer as an example [G 40]. The author used *glass capillaries* (0.2–0.45 mm i.d.) in which he prepared separation and spacer gels according to the method of Ornstein and Davis (p. 53). Recently he scaled down his technique to separate picogram quantities of proteins in capillaries of 25 μm i. d. [G 42a].

The following steps are carried out under the stereomicroscope. Glass capillaries (1–5 μl) ("Microcaps" of the firm 7) are broken into sections of 15 mm length and perpendicularly mounted in the ring-shaped ends of 1 mm thick glass rods with the aid of paraffin. The glass rods are sealed with picein into 3 mm sections of a glass tube; these tube sections are inserted into the instrument holder of a De Fonbrune micromanipulator. The capillaries are completely filled by immersion into separation gel solution. The upper part of the solution is removed by suction with a fine wick of celluloseacetate film. The inside wall of the capillary will remain wet. This is important, because the gel solution must be overlayered very slowly with water to produce a sharp interface (p. 56). This is accomplished by a capillary with a slightly curved tip attached to a tripod with a rack and pinion drive. The capillary is filled with water until the water drains freely and sufficiently slowly from it when the tip contacts the film of gel solution on the inside capillary walls. The water layer is again removed with a celluloseacetate film wick. The spacer gel solution is dispensed from a micropipette with a bent tip having an orifice of 30–50 μ diameter. This micropipette is attached to a second micromanipulator. The space over the separation gel is first rinsed with spacer gel solution.

As in the original procedure, the volume of spacer gel depends on the volume of sample solution. Usually, however, the described gel dimensions are sufficient for the electrophoresis of protein quantities with a limited number of cells in native solution. For example, the separation of proteins from the salivary gland secretion of *Chironomus* requires 0.2 μl spacer gel solution per capillary of 0.4 mm i.d. The secretion is adsorbed on a celluloseacetate film, the adsorbate is introduced into the capillary until it contacts the spacer gel and 0.1 μl spacer gel solution is added to prevent protein back-diffusion into the electrode buffer. The capillary is then twisted into a short piece of silicone tubing. The tubing with capillary is finally transferred into a glass tube (5 mm i.d.) which is installed in an ordinary disc electrophoresis apparatus. During protein concentration, the voltage gradient is first held at 10 V/cm, but as soon as the proteins have reached the separation gel, it is increased to 40 V/cm. The electrophoresis time for serum proteins, for example, amounts to 10 min. Immediately after electrophoresis, the gels are frozen in dry ice to prevent diffusion of the separated substances and are expelled from the capillary by a jet of water or by a closely fitting stainless steel wire. Protein staining with Procion (see page 72) for several min., destaining by the leaching method.

Similarly, HYDÉN et al. [H 49] used capillaries of 0.2 mm diameter for protein quantities of 10^{-7} to 10^{-9} g and analyzed the Amido Black-stained gels with a microdensitometer which, due to its high amplifier capacity, permitted the detection of *nanogram quantities*. The method of KOENIG and BRATTGARD [K 26] was used by these authors to determine low tritium activities (p. 95); they described an *interferometric* method for protein quantitation. See also [H 50].

After preparing the separation gels in glass capillaries (55 mm length, 0.215 mm i.d., 2 μl capacity), 0.15 μl of a Sephadex G 25 Superfine suspension in phosphate buffer is applied, the particles are allowed to settle and the excess of buffer is removed. Subsequently, 0.85 μl water is added in which 20–30 isolated cells are homogenized with a wire loop (stainless steel, 28 μ gauge) by the method of EICHNER [E 3]. The loop is held by a Teflon tube clamped into a dental drill and is driven at 12,000 rpm. Electrophoresis takes place at 1.5 μA for 3–4 hrs. The gel is then pushed out of the capillary with a steel wire, is stained and destained. For the determination of the relative mobilities, 25 μg fluorescein is added per gel which migrates with the buffer front. Its location can be visualized under a UV lamp. The detection limit of the microdensitometer method is about 3.6×10^{-11} g serum albumin, while the optimum quantity of cell protein amounts to 10^{-7} to 10^{-8} g.

NEUHOFF [N 27] modified the latter method in order to achieve optimal fractionation of low molecular weight brain proteins in 25 % gels containing 0.5 % hydantoin which markedly improved band sharpening during protein separation (Fig. 46).

1 % Triton X-100 is added to the gel solution for easy extruding the electrophoresed gels from the 2–10 μl capillary tubes. Instead of a Sephadex Superfine spacer gel on top of the separation gel, a 5 % polyacrylamide gel is used since Sephadex does not provide a tight contact with the surface of the separation gel and gives rise to leakage of protein into the capillary slit between the glass wall and the gel. For the separation of relatively small proteins it was found advisable to use undiluted electrode buffer and constant voltage (60 V) rather than constant current. A special power supply is used [N 29] (see p. 41). See also [N 28, N 30–34] for application of the method. The gels are scanned with a microdensitometer from firm 17. The whole procedure including scanning consumes only 1–1.5 hrs. Less than 10^{-9} g of albumin can be detected in 5 μl-gels. For micro-disc electrophoresis combined with micro-enzyme and immunoanalysis see page 114.

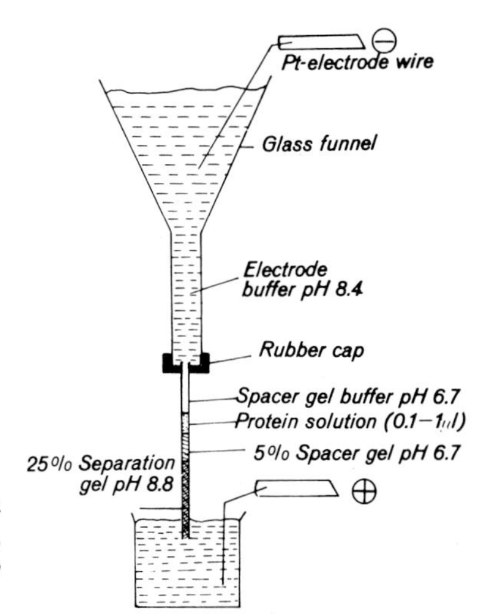

Fig. 46: Micro-disc electrophoresis in a 5 μl glass capillary tube (length 32.5 mm, internal diameter 450 μm) according to NEUHOFF [N 27].

Other workers found gel diameters intermediate between those of the microgels (0.2–0.5 mm i.d.) and the standard gels (5 mm i.d.) adequate to meet their requirements. Thus HEMMINGSEN and other [H 18] and JIRKA [J 6] used 2 mm gels to study aqueous humour and sweat proteins, respectively. KRAUSE and RAUNIO [K 35] and FELGENHAUER [F 12, F 14] analyzed serum proteins, RAPOLA and KOSKIMIES [R 7] LDH isozymes from mouse eggs on 1 mm columns. PALVA and RAUNIO [P 1–2] reported on the separation of human peri- and endolymph proteins on 0.8 mm gels. BURR et al. [B 62a] described a modification by which relatively large loading volumes, hence very dilute protein solutions can be applied.

MATIOLI and NIEWISCH [M 21] developed a remarkable procedure for the separation of *picogram quantities* of hemoglobin A_2, A, C, S and J from *single* erythrocytes. Since the average quantity of hemoglobin (Hb) in an erythrocyte amounts to about 3×10^{-11} g and the average Hb-A_2 content represents 2.5 % of the total Hb, this method permits the detection of about 7×10^{-13} g of Hb-A_2.

A *gel fiber* in which the erythrocytes are gelated serves as the electrophoretic matrix. Protein detection is accomplished by microscopy with monochromatic light ($\lambda = 418$ nm), because Co-Hb has an absorption maximum at this wavelength.

Fig. 47 shows the electrophoretic separation of the hemoglobins of single erythrocytes. In some cases, the stroma is no longer visible as a result of lysis (B a, C a + b). The method opens possibilities for the quantitative analysis of γ-, β- and δ-hemoglobin chains of individual cells as well as for a more detailed elucidation of the genetic activity of erythroblasts.

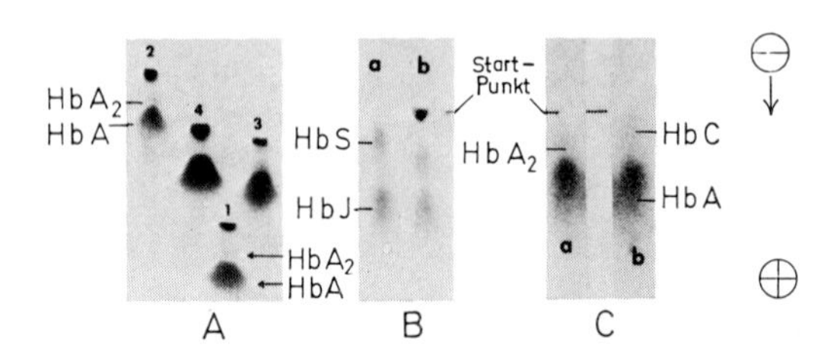

Fig. 47: – Ultramicro electrophoresis of hemoglobins of single erythrocytes. From: MATIOLI and NIEWISCH [M 21].

(A) Erythrocytes of normal adults. Hb-A_2 visible only in erythrocytes 1 and 2. 44 % rel. humidity in the fiber. (600 ×.)

(B) Erythrocytes of a patient with Hb S and Hb J. 65 % rel. humidity in the fiber. (600 ×.)

(C) Erythrocytes of (a) a normal adult and (b) a patient with Hb C on two different fibers. 65 % rel. humidity in the fiber. (1000 ×.)

3.2. Micro-Disc Electrophoresis combined with Micro-Enzyme and Immunoanalysis

NEUHOFF and LEZIUS [N 28] elegantly *combined micro-disc electrophoresis and micro-enzyme analysis* to identify small amounts of enzymes in 5 µl gels. Two methods for the identification of *DNA polymerase* among other proteins of crude preparations were developed. Both tests could be equally well applied for the assay of DNA-dependent *RNA polymerase*.

1. The *binding test* as a "negative" method in which DNA polymerase forms a complex with poly d(A–T) in the presence of Mg^{2+} ions, dATP, and dTTP. This complex of rather high molecular weight cannot enter the separation gel and remains in the interphase between spacer and separation gel. Consequently, the appropriate zones are missing after staining with Amido Black 10 B in acetic acid (Fig. 48).

2. The *polymerization test* as a "positive" method in which the separation gel contains a minute amount of poly d(A–T) primer, sufficient to induce an extensive synthesis of poly d(A–T) by DNA polymerase in the gel when incubated with dATP and dTTP in the presence of Mg^{2+} ions. The newly synthesized poly d(A–T) is stained with pyronine and can be estimated with a sensitive microdensitometer (Fig. 49).

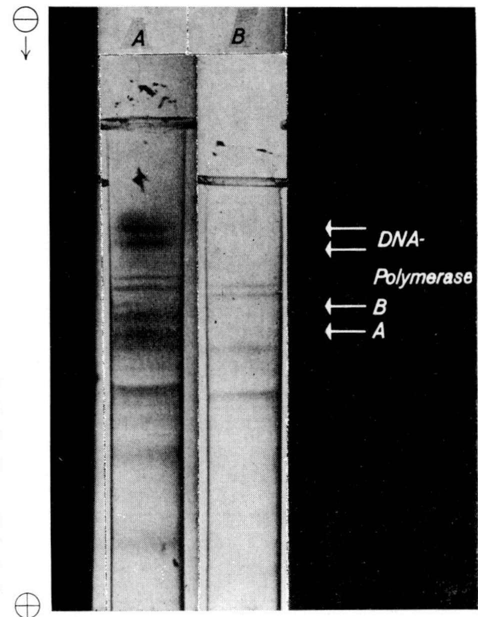

Fig. 48: Identification of DNA polymerases in 25 % microgels (5 μl) by the binding test (see page 114). From: NEUHOFF and LEZIUS [N 28].

A: Crude DNA polymerase fraction VII, B: same protein quantity (0.5 μg) plus primer poly d(A–T), dATP, dTTP, Mg^{2+} and phosphate buffer. Amido Black staining, Magnification 38 fold.

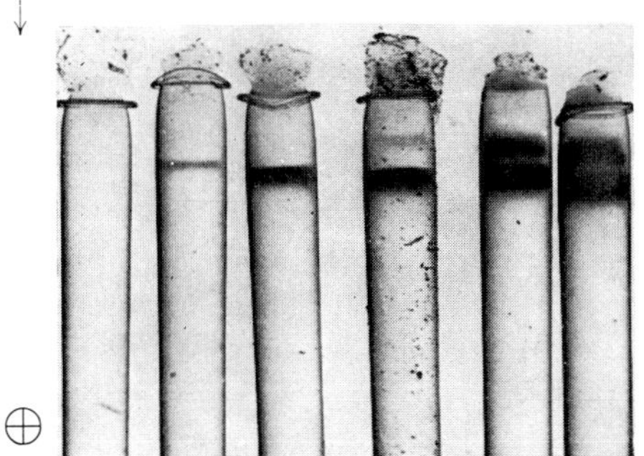

Fig. 49: Identification of DNA-polymerases in 25 % microgels (5 μl) by the polymerization test (see page 114). From: NEUHOFF and LEZIUS [N 28].

All gels contained poly d(A–T) and were loaded with crude DNA polymerase fraction VII. Incubation in A for 10, B for 20, C for 30, D for 40, E for 50 and F for 60 min. at 37° C with dATP, dTTP, Mg^{2+} and phosphate buffer. Pyronine/acetic acid staining. Magnification 20 fold.

Similarly, NEUHOFF and SCHILL [N 30] *combined micro-disc electrophoresis and micro-agar immunoprecipitation* to identify small quantities of protein fractions.

The proteins are fractioned in 5 μl-gels (see p. 113), the distinct stained protein bands separated by slicing and eluted electrophoretically from the gel slices into buffer liquid (Fig. 50). The eluted proteins are subjected to micro immunoprecipitation using an agar layer on a microscope slide. 0.02 μg serum albumin can be detected by its precipitation line following 30 min. immunodiffusion. It is noteworthy that even bromophenolblue and Amido Black stained and acetic acid treated proteins maintain their antigeneity for the immunoreaction, i.e. the molecular groups responsible for the precipitation are, at least, not significantly altered by acetic acid and the dyestuff. This method is particularly suitable for the immunological identification of small amounts of proteins and should prove valuable for forensic analysis. See also [N 31–34] for application of the method.

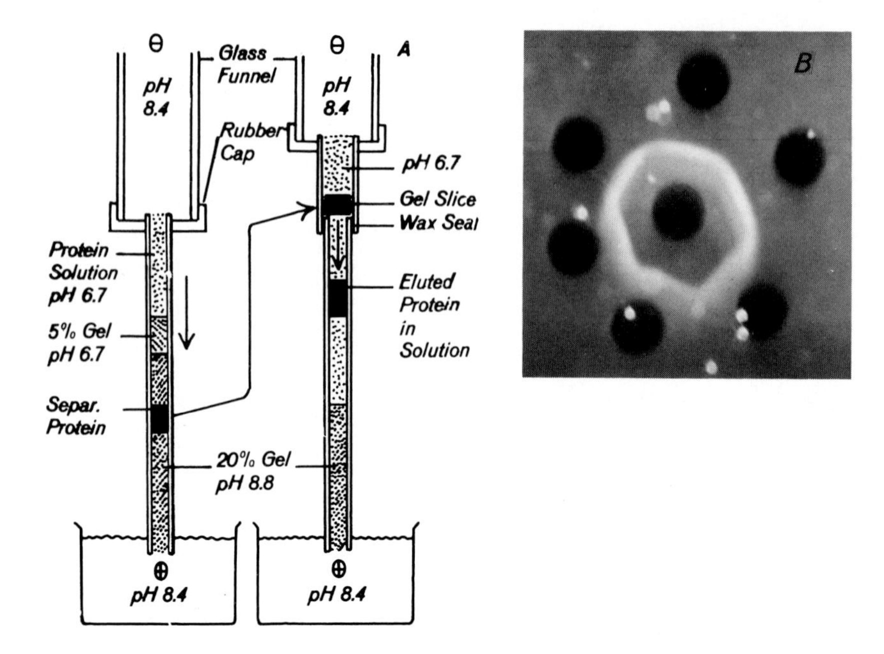

Fig. 50: Combined micro-disc electrophoresis of human serum proteins and micro-immuno- precipitation of isolated albumin against rabbit anti-albumin serum. From: NEUHOFF and SCHILL [N 30].

A: Electrophoresis of 1 μl of serum (diluted 1 : 2000 with 0.15M NaCl) at 80 V. Gels stained for 5 sec. with 0.25 % bromophenolblue in 7.5 % acetic acid, washed immediately in 0.15 M NaCl and segmented. Albumin fraction transferred to second capillary tube and electrophoresed into Tris-phosphate buffer pH 6.7.

B: Albumin solution (1 μl) filled into holes of agar layer (1 % purified agar in 0.037 M potassiumphosphate buffer pH 7.1, 1 mm thick on microscope slide). Magnification 10 fold.

4. PREPARATIVE POLYACRYLAMIDE GEL ELECTROPHORESIS

4.1. General Comments

The excellent results obtained with analytical disc electrophoresis on a microscale soon led to the wish to achieve similar resolutions on a preparative milligram scale. Designs of equipment for the separation and preparation of fractions of maximum homogeneity have not been lacking. Yet, all reflect the technical problems involved in the application of microscale technique to macro-conditions.

Preparative electrophoresis must satisfy two conditions if it is to assert itself as a productive method: first, it must have the capability to resolve large quantities of substance that can be further analyzed by other techniques; secondly, it should afford a quantitative recovery of the separated components in suitable form as far as possible. Thus, the ideal recovery process should not cause: (1) any mixing of separated zones, (2) loss of material, (3) dilution of zones, or (4) denaturation [H 28]. In the design of such equipment, the following influences on productive output should be taken into account [J 18]:

Loadability (a function of the gel cross section),
Mechanical stabilization of the support material,
Dissipation of ohmic heating,
Maintenance of a uniform electrical field geometry,
Maintenance of a hydrostatic equilibrium,
Elution method,
Duration of separation.

In addition, to achieve optimal separation conditions the following variables should be properly chosen [G 24]:

PH and ionic strength of gel buffer, electrode buffer and elution buffer,
Gel concentration,
Column height,
Amount of substances to be separated,
Voltage gradient,
Flow rate and fraction volume of elution buffer,
Operating temperature.

Usually, preparative gel electrophoresis will be used for further purification of prepurified crude extracts, i.e. it will not represent the first purification step. Otherwise, precipitates of material in the gel might plug the pores and cause smearing of the separations when they are redissolved. Moreover, with mixtures of molecules of very different molecular weight, one preparative run will hardly lead to the desired result, but rather only two or more electrophoresis runs in different gel systems will achieve this end (see page 19). Thus group separation will generally precede fine fractionation.

In most cases, it is advisible to determine the optimum separation conditions, particularly the amounts of material per cross sectional area of the gel, by the analytical method and to apply these to preparative electrophoresis.

4.1.1. Gel Preparation

In principle, the gels are prepared by the same carefully controlled operations as outlined for the analytical procedure. In particular, the following rules should be observed. Before adding the persulfate-solution, the gel solution should be degassed by a water pump and a magnetic stirrer (about 10 min.). Then the gel solution is immediately poured into the gel mold taking care not to entrap bubbles. The gel solution is quickly but thoroughly overlaid with water: lower the end of a plastic tubing fitted to a water filled syringe so that it rests just above the liquid surface. Allow the water to flow out as gently and evenly as possible, raising the tubing as the water layer grows. Avoid any mixing of the gel solution and the water overlaid. Allow the solution to stand for at least one hour unmoved and untouched (some investigators leave it overnight). Pre-electrophoresis is recommended for at least one hour under the same conditions (cooling buffer) as used for the electrophoresis later on. The sample solution may contain 5 % sucrose or 10 % glycerol to increase its density. Under the conditions to be selected the rate of elution relative to voltage and gel length merits particular attention.

4.1.2. Sample Loading

If the sample can tolerate polymerization the use of sample gel is recommended: thus upward loss by convection or reverse migration are minimized. Sample in dense solutions are still subject to convection, which may result in tailing effects. Direct loading of the sample (without dense solution) may cause precipitation.

4.1.3. Artifacts mainly arising in Preparative Disc Electrophoresis

Canalco [C 7] has compiled, after extensive studies, a number of problems peculiar for preparative disc electrophoresis and has made some useful suggestions to solve them.

1. Precipitation of material may contaminate a whole run since the material fallen behind the migrating stack can redissolve progressively.

2. Old chemicals may contain impurities. Gel swelling is directly related to acrylamide impurities.
3. Too much crosslinker can cause gel shrinkage which leads to pull-away from the walls.
4. Riboflavin forms an immobile barrier at the elution slit breaking up the stack as it reaches this point. Replace this catalyst by persulfate.
5. Too high ionic strength may cause extensive ohmic heating. It is recommended to keep the salt content lower than in analytical gels; thus less current is required.
6. Differences of the ionic strength of the gel and the elution buffer can result in gel swelling or shrinking due to the Donnan equilibrium effect. Diffusion and mixing of fractions may occur. Keep the salt contents equal. Too high salt content of the sample may prevent polymerization of sample gel; in these cases the samples should be dialyzed.
7. Overconcentrated proteins tend to precipitate.
8. Too high current may cause precipitation and denaturation of materials, distortion of bands, detachment of the gel from the wall etc.

4.2. Methods of Preparative Polyacrylamide Gel Electrophoresis

4.2.1. Extraction of Gel Fractions

As in the analytical method, the materials are subjected to electrophoresis until a sufficient separation is obtained and localized by staining of a reference strip (p. 95). The gel segments containing material are then cut out and eluted by one of the following two methods.

The apparatus described on p. 36 can be used for electrophoresis. SULITZEANU et al. [S 80–81], for example, use Plexiglas gel tubes of 28 mm i.d. and 80–105 mm length. See also [B 45]. However problems may arise with the use of rather thick gels; they are similar to those of high density gels (p. 56).

4.2.1.1. Elution by Diffusion

The gel fractions are homogenized mechanically or by ultrasonics and eluted exhaustively with water or buffer solution. The eluates are separated from the gel particles by centrifugation or filtration. In principle, this corresponds to the method described on p. 95, to which reference is also made for pertinent comments. As an example see [L 24]. Main disadvantage of the method: Low yields are likely to occur.

4.2.1.2. Elution by Electrophoresis

The individual sample fractions are separated in the spacer gel according to their mobilities and concentrated in sharp zones; the width of these zones is proportional to the particular component quantity in the sample solution.

Therefore, when corresponding gel fractions of several separation gels are combined after electrophoresis under identical conditions, are polymerized with spacer gel solution, and the new fraction-spacer gel is subjected to electrophoresis until the component is swept from the gel (free flow electrophoresis), the fraction can be collected in purified and concentrated form in a dialysis bag [L 24]. With the use of long columns (for example, 100 cm), it is also possible to concentrate large amounts of material to form relatively thick zones in the spacer gels which contain the components in highly purified form; they can then be recovered in purified and concentrated form by free flow electrophoresis in the sequence of their mobilities [O 9].

A simple method consists of placing the disks of a separation gel firmly on a porous frit in a tube, to cover them with a layer of buffer solution and to subject the materials to electrophoresis, allowing them to collect in an attached dialysis bag [S 81]. See Fig. 50 A for a similar method.

Schwabe [S 27] developed a collecting chamber for the recovery of proteins from standard gels. In this system, the fractions are eluted into the chamber, which is filled with insoluble starch, by electrophoresis perpendicular to the separation direction; the starch is fractionated and the fractions are extracted. The total recovery amounts to about 30 %.

4.2.2. Continuous Separation and Elution by Electrophoresis

With the use of this method, the separated components migrate to the lower end of the gel where they are entrained by a continuous flow of buffer and collected individually.

4.2.2.1. Micropreparative Devices

Micropreparative elution devices have been developed for recovery of microgram quantities of proteins and nucleic acids. A simple elution attach-

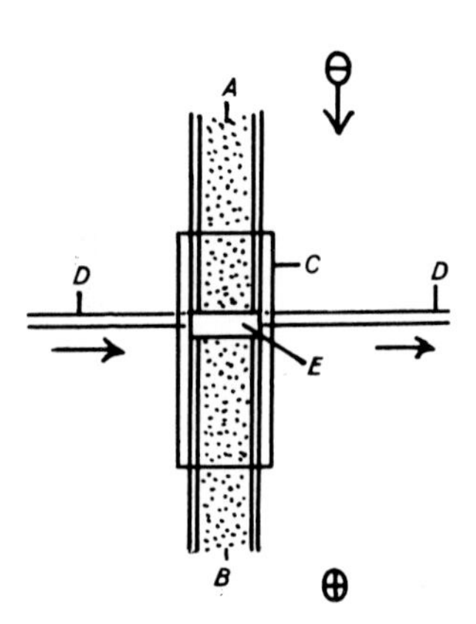

Fig. 51. Elution attachment for micropreparative gel electrophoresis.

A: separating gel, B: retaining gel, C: joining Tygon tubing (6.35 mm i.d.) with D: inserted PE-50 polyethylene tubings for buffer in- and outlet, E: elution chamber of 2 mm height. From Mann and Huang [M 8a].

ment for use with conventional disc electrophoresis apparatus was designed by MANN and HUANG [M 8a] (Fig. 51), whereas the device by RAY et al. [R 9a] is an efficient microversion of apparatus types described below. An ultramicro device is shown in Fig. 50 A.

4.2.2.2. Preparative Apparatus for Milligram Sample Quantities

A great number of instruments for this purpose have been designed in research laboratories and their use reported [for example, A 25, D 37, G 24, H 26, H 28, H 36, J 18, M 2, R 31, R 60, S 54, S 75].

Critical points of most apparatus developed to perform this method concern: (1) Cooling of the gel: A temperature gradient cannot be completely avoided but should be minimized. It may lead to curvature of the separated bands and reduced resolution. (2) Volume of elution chamber: It should be kept as small as possible to provide a high turnover of the elution buffer. (3) Support of the gel: A hydrostatic equilibrium should be sustained. (4) Membrane of the elution chamber: Adsorption of sample material to this membrane should be prevented.

DUESBERG and RUECKERT [D 37] gave considerable data on useful buffer and gel systems (Tables 13 and 14).

Table 13: Composition of electrode and elution buffer solutions according to DUESBERG and RUECKERT [D 37].

Component	Electrode buffer				Elution buffer		
Tris (g)	6.0	—	—	—	52.0	—	—
Glycine (g)	28.8	—	28.1	—	—	—	—
β-Alanine (g)	—	31.2	—	—	—	—	—
Acetic acid (ml)	—	8.0	4.0	60.0	14.0	19.0	100.0
Ammonium acetate (g)	—	—	—	42.0	—	14.0	14.0
10 M urea (ml)	—	—	—	—	800.0	800.0	800.0
pH	8.5	4.6	3.8	4.3	8.5	4.9	4.3
Ionic strength μ	0.11	0.115	0.105	0.475	0.16	0.155	0.43

Quantities per liter of aqueous solution.

Table 13 shows the compositions of the electrode and elution buffer solutions and Table 14 those of the different gel systems. In principle, the ascending counterions of the elution buffer and the gel buffer should be the same. It is of advantage to select volatile components for the elution buffer in order to dispense with the dialysis of the eluate. The *ionic strength* of the elution buffer should be higher than that of the gel buffer to allow the formation of sharp zones when the proteins enter the collecting chamber (p. 25) and to maintain a minimum adsorption on the frit or membrane. – It is advisable to use deionized urea. Deionization: 1 l 10 M urea is agitated in 30–100 ml "mixed-bed resin" (available from firm 3) for 1 hr, is filtered and stored at 3–6° C.

Table 14: Composition of buffer systems for separation and spacer gels according to DUESBERG and RUECKERT [D 37].

Component	pH 9.9 system		pH 4.5 system		pH 3.8 system	
	SG	SpG	SG	SpG	SG	SpG
10 N HCL (ml)	12.0	12.0	–	–	–	
Tris (g)	80.0	72.6	–	–	–	
CH_3COOK (g)	–	–	30.0	30.0	30.0	
CH_3COOH (ml)	–	–	120.0	60.0	420.0	see
TEMED (ml)	5.8	11.5	25.0	3.0	25.0	legend
10 M urea (ml)	800.0	800.0	800.0	800.0	800.0	
pH after 10-fold dilution	9.5	9.5	4.8	4.9	4.3	
Ionic strength μ	0.233	0.255	0.725	0.3751	1.7	

Quantities per liter of aqueous solution except for the pH 3.8 separation gel. The composition of the pH 3.8 spacer gel corresponds to that of the pH 4.5 spacer gel. SG = separation gel, SpG = spacer gel.

Composition of the gel solutions: *Separation gel*: 30 g acrylamide and 0.8 g Bis are dissolved in 80 ml 10 M urea. – *Spacer gel*: 10 g acrylamide and 2.5 g Bis are dissolved in 80 ml 10 M urea and water added to 100 ml solution. *Catalyst solution* for acid spacer gels: 0.04 % aqueous riboflavin. Catalyst solution for all other gels: 0.15 % ammonium persulfate in 8 M urea.

The following studies of DUESBERG and RUECKERT [D 37] on the influence of *pH value* and *sample quantity* on peak width and gel capacity illustrate the significance of these parameters. The isoelectric point (IP) of tobacco mosaic virus protein (TMV) is approximately at pH 4.8. As indicated by Fig. 52, sharper zones form when the pH value is shifted from 4.5 to 3.8. This results in the important conclusion that proteins can be separated more readily the greater the difference between the pH and IP values, i.e. the running pH values should be 1–4 units higher or lower than the isoelectric point. For the preparative separation of larger quantities of material (e.g. 20 mg instead of 5 mg) in particular, a running pH value which is too close to the IP has an unfavorable influence: The bands of the fractions broaden and the resolution thus becomes poorer. But even with an optimum difference, the proportion of the quantity of substance to cross-sectional area of the gel plays a role in terms of the separation effect. If this ratio exceeds a certain value which must be established for each case, we can expect the fractions to overlap; this was confirmed by the authors in a satisfactory preparative separation of proteins of identical chains but different charge (TMV and acetyl-TMV protein) in the pH 3.8 gel system.

In this connection, we should note two effects of *urea* on proteins at different pH values during electrophoresis. First, urea leads to a more or less reversible uncoiling or denaturing. Secondly, cyanate forms from urea and is in equilibrium with the

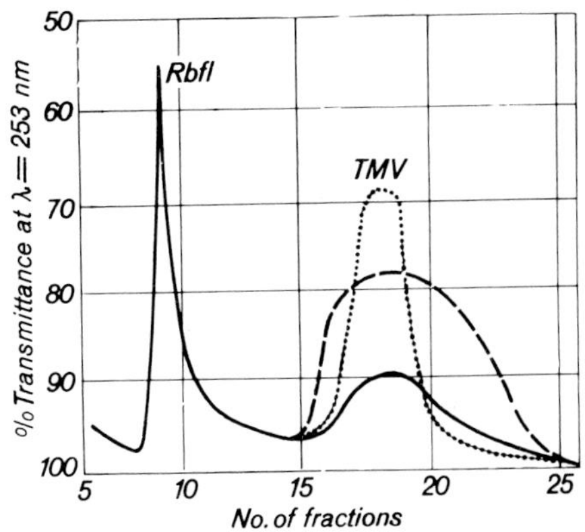

Fig. 52. – Preparative electrophoresis of tobacco mosaic virus (TMV) protein at different pH values and with different quantities. From: DUESBERG and RUECKERT [D 37].

Solid line: 5 mg protein; dashed line: 20 mg protein, pH 4.5 system (5 cm column). Separation gel: 6 cm length, 4 % acrylamide in 8 M urea, $\mu = 0.0375$; spacer gel: 1 cm length, 2.5 % acrylamide in 8 M urea, $\mu = 0.0187$. Current increased from 40 mA/100 V for fraction 1 to 95 mA/300 V for fraction 10, then constant. Flow rate of elution buffer of pH 4.9: 15 ml/h, 2.5 ml fractions. Rbfl.: riboflavin. Electrode buffer of pH 4.6.– Dotted line: 10 mg protein in the pH 3.8 system with the following changes compared to the pH 4.5 system: separation gel: $\mu = 0.12$; electrode buffer of pH 3.8 (anode), pH 4.3 (cathode). Yields (fractions 17–21): 99 % in the pH 3.8 experiment.

latter as a function of the pH and temperature [C 48]; cyanate can carbamylate the amino and sulfhydryl groups of proteins. At pH < 5.5 and > 9 such reactions are practically eliminated; at neutral pH, however, they should be suppressed by sulfhydryl reagents.

Slowly migrating components occasionally exhibit broader elution profiles than faster fractions. In these cases, one or more of the following parameters is modified: the gel concentration is decreased, the separation distance is shortened, the elution time is reduced (if possible, discontinuously according to a suggestion of JOVIN et al. [J 18]) or the voltage gradient is increased, but in that case, the increased generation of ohmic heating must be kept in mind.

JOVIN et al. [J 18] designed an apparatus to meet most of the requirements for efficient operation (Fig. 53).

The membrane holder (16) is covered with a wet dialysis membrane, inserted into the cooling jacket (17) sealed against water, and immersed into the lower electrode buffer; air bubbles are removed with curved-tip pipettes (p. 70). The collecting and elution chamber (20) over the membrane is filled with elution buffer. The cooling finger (18) is inserted into the gel holder (19). Separation gel solution is slowly introduced with cooling. After gelation, the spacer gel is polymerized over it as usual. The gel holder (with cold finger) is then installed in the cooling jacket (17); air bubbles must not be allowed to form. Sealing of the collecting chamber is difficult. The best method is to clamp the membrane between a rubber gasket and a groove in the membrane holder.

For example, JOVIN et al. [J 18] used the apparatus to separate the hemoglobins A and S and three other hemoglobins in different buffer systems with hemolysate quantities of 26 to 37 mg (volume of separation gel 50–60 ml,

Table 15: Gel systems for preparative polyacrylamide gel electrophoresis. From: JOVIN et al. [J 18]

System	Mixing ratio (v/v)	Separation gel stock solution components/100 ml		pH	Mixing ratio (v/v)	Spacer gel stock solution components/100ml		pH	Lower buffer	Upper buffer	Elution buffer
									components/1000 ml		
Tris	1	Acrylamide Bis	30.0 g 0.8 g		2	Acrylamide Bis	5.0 g 1.25 g		1 N HCl 60.0 ml Tris to pH 8.1	Gly 2.88 g Tris 0.6 g pH 8.3	As lower buffer
	1	1 N HCl Tris TEMED	24.0 ml 18.15 g 0.23 ml	8.9	1	1 M H_3PO_4 Tris	12.8 ml 2.85 g	6.9			
	2	Per	0.14 g		1	Riboflavin	2.0 mg				
Tris-urea	1	Acrylamide Bis 10 M urea	21.2 g 0.56 g 80.0 ml		2	Acrylamide Bis 10 M urea	5.0 g 1.25 g 80.0 ml		1 N HCl 60.0 ml Tris to pH 8.1	Gly 2.88 g Tris 0.6 g pH 8.3	As lower buffer
	2	1 N HCl Tris TEMED 10 M urea	12.0 ml 9.07 g 0.12 ml 80.0 ml		1	1 M H_3PO_4 Tris 10 M urea	12.8 ml 2.85 g 80.0 ml				
	1	Per 8 M urea to	0.28 g 100 ml		1	Riboflavin 8 M urea to	2.0 mg 100 ml				
N-ethyl-morpholine	7	Acrylamide Bis	60.0 g 1.6 g		2	Acrylamide Bis	5.0 g 1.25 g		EM 4.46 ml AcOH to pH 8.0	Gly 2.88 g EM to pH 8.3	EM 44.6 ml AcOH to pH 8.0
	32	1 N HCl TEMED EM to pH	7.5 ml 0.07 ml 8.9	8.9	1	1 M H_3PO_4 EM to pH	12.8 ml 6.9	6.9			
	1	Per	2.8 g		1	Riboflavin	2.0 mg				

Table 15 Gel systems for preparative polyacrylamide gel electrophoresis. From: Jovin et al. [J 18]

N-ethyl- morpho- line-urea	5	Acrylamide 54.0 g Bis 1.44 g 10 M urea to 80.0 ml		2	Acrylamide 5.0 g Bis 1.25 g 10 M urea 80.0 ml		EM 4.46 ml AcOH to pH 8.0	Gly 2.88 g EM to pH 8.0	EM 44.6 ml AcOH to pH 8.3
	33	1 N HCl 7.3 ml TEMED 0.07 ml 10 M urea to 80.0 ml EM to pH 8.9	8.9	1	1 M H_3PO_4 12.8 ml 10 M urea 80.0 ml EM to pH 6.9	6.9			
	2	Per 1.4 g 8 M urea to 100 ml			Riboflavin 22.0 mg 8 M urea to 100 ml				
Tris- stacking gel				2	Acrylamide 15.0 g Bis 0.4 g		1 N HCl 60.0 ml Tris to pH 8.1	Gly 2.88 g Tris 0.6 g pH 8.3	
				1	1 M H_3PO_4 12.8 ml Tris 2.85 g TEMED 0.23 ml	6.9			
				1	Per 0.7 g				

Bis = N,N′-methylene-bis-acrylamide
TEMED = N,N,N′,N′-tetramethylethylenediamine
Per = ammonium persulfate
EM = N-ethylmorpholine
Gly = glycine
AcOH = glacial acetic acid

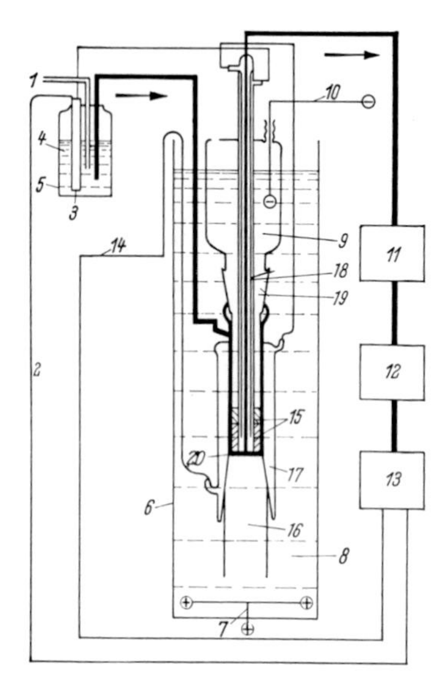

Fig. 53: – Schematic overall view of preparative electrophoresis equipment. From: JOVIN et al. [J 18].

Heavy line: flow of elution buffer.

1 air inlet of the elution buffer reservoir
2 cooling coil
3 cooling finger
4 elution buffer solution
5 elution buffer reservoir
6 lower electrode compartment
7 anode (Pt)
8 lower electrode buffer
9 upper electrode buffer
10 cathode (Pt)
11 pump for cooling finger
12 spectrophotometer with recorder
13 cooling system with circulation pump
14 cooling coil
15 upper and lower gel
16 membrane holder
17 cooling jacket
18 cooling finger
19 gel holder
20 collecting chamber

elution rate 1.2–1.0 ml/min, electrophoresis at 830–940 V and 30–56 mA). The authors developed a number of gel systems for preparative purposes (Table 15).

Such gel systems need to satisfy the following conditions: The *elution buffer* must contain the counterion; in the listed gel systems, this is a cation. The free base and weak conjugated acid of the corresponding anion should be volatile so that the eluates will not need to be dialyzed prior to freeze-drying. As noted (p. 121), the ionic strength of the elution buffer should be higher than that of the separation gel, so that on leaving the separation gel, the proteins will encouter a medium of high conductivity and low field strength and therefore cannot tend to surface adsorption. The upper limit of the ionic strength is determined by the solubilities of the components and the degree to which the protein bands are slowed during elution from the gel due to the diffusion of elution buffer into the separation gel. The pH of the elution buffer should be selected such that an optimum buffer capacity is obtained and the mobilities of the proteins migrating from the gel are decreased. In urea-containing gel systems, the elution buffer may contain no urea. The composition of the lower electrode buffer is not critical, however; it should also contain the counterion but may have a much lower concentration.

The *elution rate* depends on the electrophoretic mobilities of the sample material as well as on the separative power of the gel. It is therefore necessary to weigh the advantages of low migration velocities (minimum dilution and small volumes of fractions) against the need to transfer the contents of the elution chamber frequently enough so that the separated component zones will not overlap. It should be noted that after passing through the interface, polyacrylamide gel does not release UV-absorbing materials during a continuous elution of the individual fractions, so that stable base lines are obtained for optical recording of the fractionation. With the use of a printing recorder, the peaks can be evaluated on the basis of the following equation containing the elution rate as a function of time:

$$\text{Quantity}_i = \frac{1}{k_i l} \int_{t_1}^{t_2} A_i(t) f(t) dt$$

where quantity$_i$ = amount of component i represented by peak i; k_i = specific absorbance of the component i; t_1, t_2 = time limits for peak i; $A_i(t)$ = absorbance of component i; $f(t)$ = elution flow rate; l = length of cuvette light path.

The apparatus by JOVIN et al. [J 18] has been modified by GORDON and LOUIS [G 24] and SMITH and MOSS [S 54–55] particularly to improve the elution method (elution chamber). See also [H 36].

In collaboration with ORNSTEIN, the Canal Industrial Corporation (firm 5) designed an apparatus for preparative disc electrophoresis which is characterized by interchangeable columns of different capacity, internal cooling and externally adjustable elution chambers with variable volume (Fig. 54). The

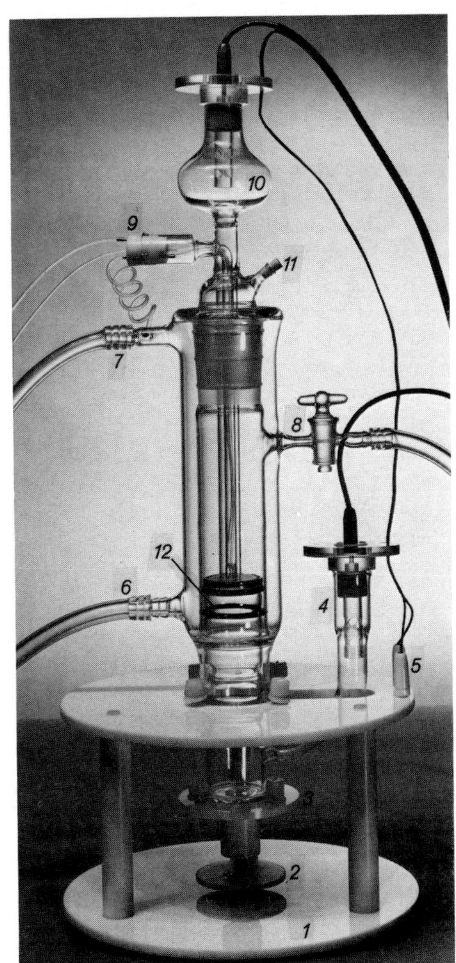

Fig. 54: Assembled preparative disc electrophoresis apparatus of CANALCO (firm 5).

Components: 1 Base assembly, 2 slit control wheel, 3 pedestal, 4 lower electrode, 5 safety interlock, 6 jacket coolant in, 7 jacket coolant out, 8 elution buffer inlet, 9 core coolant in, eluate out and core coolant out, 10 upper electrode reservoir and electrode, 11 elution chamber stopper, 12 elution chamber with slit disc at bottom.

latter is important since polyacrylamide gels exposed to low ionic strength buffers tend to swell during prolonged electrophoresis due to the Donnan equilibrium effect. This results in a convex shape of the button surface of the gel reducing the elution chamber volume. Moreover, several useful separating methods and gel systems have been developed (Table 16).

Thus, *stacking gels systems* are mainly suited for the separation of large from small molecules and for the concentration of certain groups of components which are present only in small quantities in the starting material (*group separation*). In these systems, the components are arranged according to their mobilities during concentration and are separated in this manner. We thus distinguish stacking gel systems without sieving effect (non-sieving stackers, 2–3 % gels) and those with sieving effect (sieving stackers, 3.5 % gels). The latter retain large molecules more firmly than smaller: as a result, certain groups of components can be separated and can be further fractionated by re-electrophoresis. The advantages of stacking gel systems reside in the short running times (even for large molecules) as well as in the concentrating effect and reduction of diffusion.

Separation gel systems are especially suited for the separation of molecules of similar mobility but different size, i.e. for further fractionation of components of a group that has been prepurified by a stacking gel system (*fine fractionation*). These systems can furnish electrophoretically pure fractions. Experience has shown that stacking and separation gel systems for preparative purposes should contain less cross-linking agent than analytical gels (0.075 % instead of 0.2 % Bis). Separation gels containing 15–20 % acrylamide are suited for the separation of materials with molecular weights (MW) of $1–4 \cdot 10^4$, with 10–15 % for MWs of $4 \cdot 10^4$ to $1 \cdot 10^5$, 5–10 % for MWs of $1–3 \cdot 10^5$, 5 % for MWs of $3–5 \cdot 10^5$, 2–5 % for MWs of $> 5 \cdot 10^5$.

It should be mentioned, that in preparative gel electrophoresis it is advisable to start with low current, since high current will cause sample precipitation, convection in sucrose or density solutions loads, or both. Also, stacking gel systems cannot take as much current as separation systems. Riboflavin as catalyst should not be used in preparative disc separation gels, but may be taken for sample and some stacking gels.

The problem of synchronous control of the elution rate and recording was solved by Strauch [S 75] with an automatic system (Fig. 55).

The separation column and elution chamber of the electrophoresis assembly are immersed in cooled electrode buffer solution. Pump 2 with its drive provides for the cycling of electrode buffer solution in the two electrode reservoirs. Pump 1 transports elution buffer solution from the buffer reservoir throught the elution chamber. Its delivery during separation is synchronized with the mobilities of the consecutive fractions eluted from the gel by control of the elution process. Consequently, the concentration ratios prevailing in the gel are also maintained in the eluate and dilution of the slower fractions is minimized. The eluate flows through a flow photometer while a recorder continuously records the concentrations. The control module of the recording system regulates the paper speed and is synchronized with the elution control. Thus, the areas under the curve peaks become directly proportional to the fraction concentrations. The elution curves assume the shape of absorbance curves as in direct densitometry of stained gel bands. The eluate is collected in the fraction collector which is controlled by the recording instrument

Table 16: Gel systems for preparative disc electrophoresis according to CANALCO [C 7].

Gel system	Separation gel solutions			Spacer (stacking) gel solutions				Electrode buffer solution		Elution buffer solution
	No.	Components per 100 ml solution	Mixing ratio (v/v)	No.	Components per 100 ml solution	pH	Mixing ratio (v/v)	Components per 1000 ml solution Cathode	Anode	Mixing ratio (v/v)
pH 8.9–10% separation gel	43	Acrylamide 40.0 g Bis 0.12 g	1 p. No. 1 2 p. No. 43 1 p. H_2O 4 p. No. 3	44	Acrylamide 14.0 g Bis 0.25 g		1 p. No. 4a 2 p. No. 44 1 p. No. 6 4 p. H_2O	Tris 3.0 g Glycine 14.4 g pH 8.3		1 p. No. 1 7 p. H_2O
pH 6.9–3.5% stacking gel				45	1 N HCl 4.80 ml Imidazole 0.598 g TEMED 0.12 ml	6.9	1 p. No. 45 2 p. No. 46 1 p. No. 47 4 p. H_2O 8 p. No. 3	Tris 3.0 g Glycine 14.4 g pH 8.3	Imidazole 3.0 g Glycine 14.4 g	1 p. No. 45 16 p. H_2O
				46	Acrylamide 28.0 g					
				47	Bis 1.7 g					

Composition of solutions 1, 3, 4a and 6, see Table 5. Sample gel for the pH 8.9–10% separation gel: In the mixing ratio of spacer gel, 4 p. H_2O are replaced by 4 p. sample solution. Sample gel for the pH 6.9–3.5% stacking gel: In the mixing ratio of stacking gel, 4 p. H_2O and 8 p. No. 3 are replaced by 2 p. No. 6 and 10 p. sample solution. Abbreviations: See p. 43. – For a 5% separation gel: Instead of solution No. 43, use the following solution: 20.0 g acrylamide + 0.48 g Bis; 2.5% separation gel: 10.0 g acrylamide + 0.48 g Bis; 15% separation gel: mixing ratio of the separation gel: 1 p. No. 1, 3 p. No. 43, 4 p. No. 3. For a 7% stacking gel: In the mixing ratio, 2 p. No. 46 are replaced by 4 p. No. 46 and 4 p. H_2O by 2 p. H_2O.

through the automatic system of the fraction collector so that each fraction will discharge into a single container. The process has a capacity for separating and analyzing a sample quantity of up to 150 mg in one run.

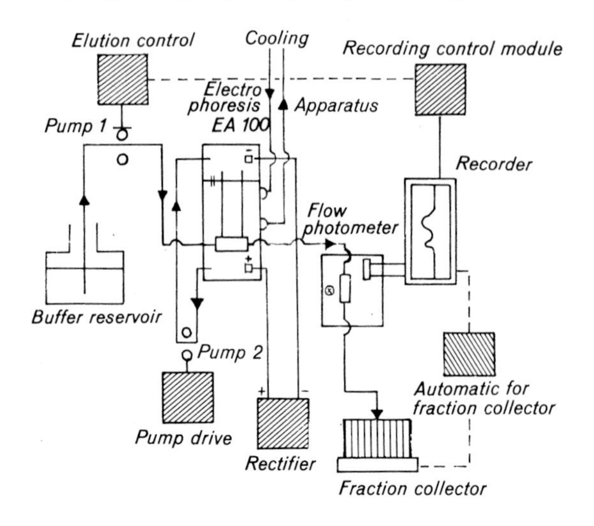

Fig. 55. Block diagram of the preparative disc electrophoresis assembly with control units for continuous elution and recording by Strauch [S 75].

For large scale separations (0.05–1.0 g of a multicomponent sample) Hjertén et al. [H 28] designed a machine with the following main features.

Cooling tubes inside the gel chamber of rectangular cross section providing efficient, uniform cooling and temperature control and a elution chamber filled with 4 % agarose gel spheres stabilizing against convection of the elution buffer. Thus membranes and their disadvantages are avoided: Contact of separated proteins to membranes may result in (1) denaturation, due to the extreme pH values at the membrane (Bethe-Toropoff effect, see [H 36]), and (2) in adsorption.

Manufacturers of equipment for preparative polyacrylamide gel electrophoresis are e. g. the firms 4, 5, 19, 27, 31, 32, 34. Only Canalco (firm 5) is licensed in the USA under the Ornstein patent to manufacture equipment for preparative disc electrophoresis.

4.2.3. Discontinuous Separation and Elution by Electrophoresis

Instead of using a continuous elution system to collect the fractions as they emerge from the gel column which may lead to high dilution of the fractions, Schenkein et al. [S 10] collect the fractions into a electrolyte cup into which the gel column dips. *Periodically* electrophoresis is stopped, the cup emptied into a fraction collector, and its electrolyte solution renewed (Fig. 56).

These operations are controlled on the basis of a five-switch recycling device. Each fraction contains the material emerging from the column while the collecting electrolyte is in place. There is little hydrostatic flow through the gel. Schenk-

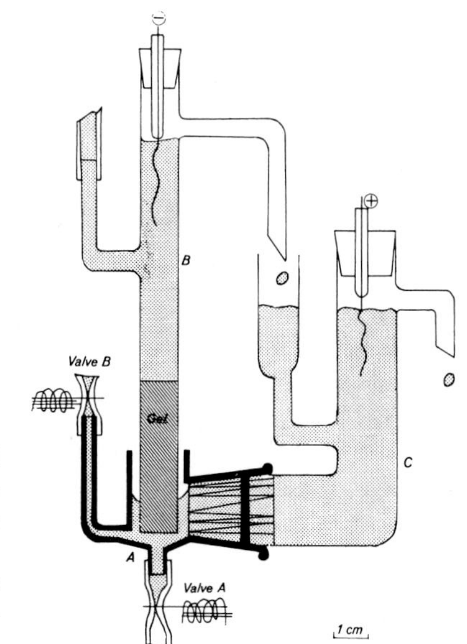

Fig. 56. Apparatus for preparative gel electrophoresis by the discontinuous separation and elution method. From: SCHENKEIN et al. [S 10].

Dotted areas are volumes filled with buffer, crosshatched areas, filled with gel. A: collecting cup, B: gel tube, C: electrode reservoir. B and C may be drained separately.

EIN et al. [S 10] also noted some *holdup* of fractions at the lower surface of the gel probably due to heterogeneous polymerization of the gel surface since the effect could be greatly reduced by cutting back 1.5 cm of the preformed gel.

BROWNSTONE [B 59] described an apparatus capable of separating up to 500 mg of protein in a gel cylinder, 4–6 cm high and 9 cm wide, surrounded by continuously circulated buffer kept at 5° C, 40–45 Watts being not exceeded. An *intermittent* collection system, by automatically controlling pump, fraction collector, recorder and power supply operation, allows a cut down of the effective flow rate of the elution buffer to only about 5 ml/hr. for a 9 cm diameter gel. Thus, slow moving components are not excessively diluted. By casting the gel in a separate mold, the dimensions of the gel are easily controlled.

5. GEL ISOELECTRIC FOCUSING

5.1. General Methods

Polyacrylamide gel has proved to be a particularly suitable anticonvection medium for *isoelectric focusing* in order to stabilize the pH gradient [A 46, C 13, C 15–16, D 2, F 9, L 9, M 1, R 41, W 34–35]. Microanalytical procedures using conventional disc electrophoresis equipment have been reported [e.g. C 13, C 15, R 41, W 34–35] (Fig. 57). Gel electrofocusing can be combined with gel electrophoresis for *two-dimensional analysis* [D 2a, M 1, W 35] as well as immunochemical techniques to yield *immunoelectrofocusing* [C 14–16, C 11a, R 41]. Three variations of the latter method have been described by CATSIMPOOLAS [C 16], a fourth introduced by CARREL [C 11a].

(1) *Disc immunoelectrofocusing*: following electrofocusing the whole polyacrylamide gel is embedded in agar gel to allow immunodiffusion against antisera (Fig. 58 A). (2) *Sectional immunoelectrofocusing* [C 15]: following electrofocusing the polyacrylamide gel is sliced and the sections subjected to immunodiffusion in agar

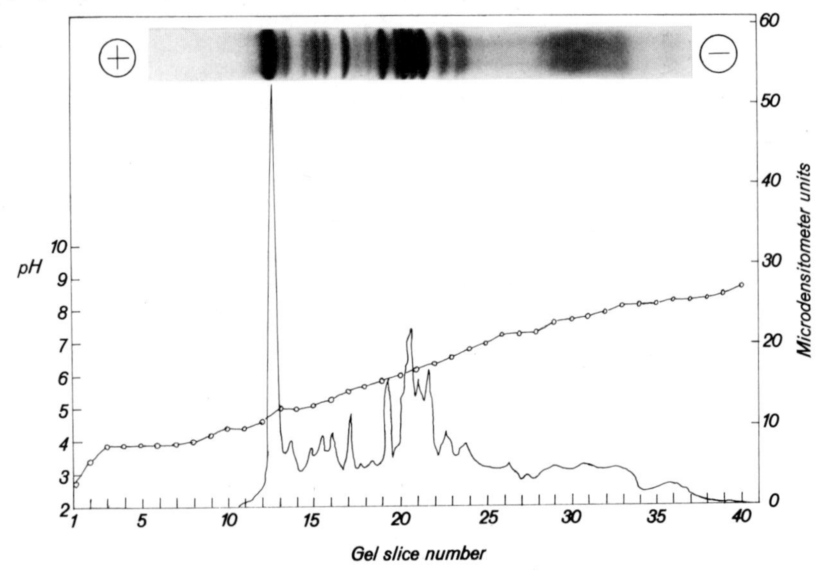

Fig. 57. Isoelectric focusing in a polyacrylamide gel column of soy bean whey proteins in the pH range between 3 and 10. From: CATSIMPOOLAS [C 13].

The densitometer tracing of the stained bands was obtained with a model F microdensitometer from firm 5. Open circles represent the pH gradient along the gel after electrofocusing.

gel (Fig. 58 B). (3) *Agarose immunoelectrofocusing* [C 14, R 41]: electrofocusing is directly performed in agarose gel on microscopic slides followed by immunodiffusion. (4) *Specific absorbent immunoelectrofocusing* [C 11a]: Following electrofocusing the gel is incubated with absorbed antiserum and stained after elution of non-precipitated proteins.

The two-dimensional combination technique of electrophoresis and electrofocusing in polyacrylamide gel can be used to discriminate proteins differing in size and charge. Such a *protein mapping* was applied for enzyme analysis (Fig. 59) by MACKO and STEGEMANN [M 1] and for human serum proteins by DOMSCHKE et al. [D 27a]. Recently GROSSBACH [G 42a] succeeded to electrofocus 10^{-10} g quantities of proteins in glass capillaries of 0.2 mm i.d. having less than 1 µl volume (*microisoelectric focusing*).

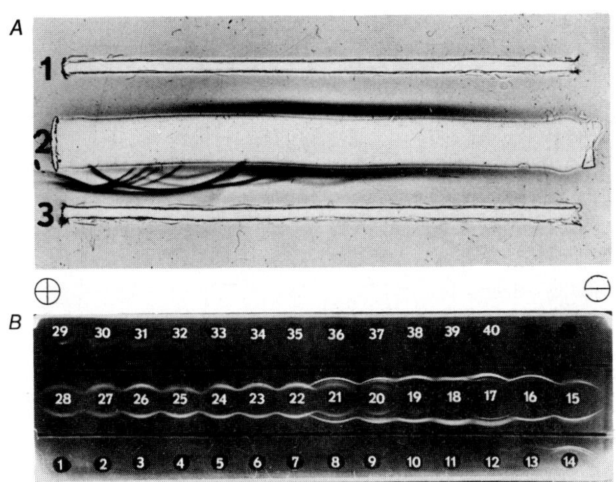

Fig. 58. (A) Disc immunoelectrofocusing of rabbit serum proteins in the pH range 5–8: (1) goat antiserum to rabbit γ-globulin, (2) electrofocused gel column, (3) goat antiserum to rabbit serum. (B) Sectional immunoelectrofocusing of rabbit serum proteins electrofocused between pH 3 and 10 and developed with goat anti-rabbit serum. From: CATSIMPOOLAS [C 15–16].

5.2. Procedures and Instructions

Table 17 lists several useful gel systems and running conditions for gel isoelectric focusing.

5.2.1. Sample Application

The sample is dissolved in the volume parts of water of the working solution (Table 17). This mixture is degassed for about 1 min. For standard gel columns (i.d. 5, length 65 mm) protein quantities between 30 and 300 µg

Table 17: Gel Isoelectro Focusing Systems

P.: Volume part (for mixing ratio of stock solutions), Ampholine: LKB Ampholine Carrier Ampholytes, 40% solution, availabe from firm 19.

Gel Isoelectro-focusing System	Stock solution No.	Components per 100 ml aqueous solution	Working gel solutions: Mixing ratio of stock solutions		Electrode solutions		Running conditions
			Photopolymerization	Chemical polymerization	Upper bath: Anode ⊕	Lower bath: Cathode ⊖	
No. 1 acc. to WRIGLEY [W 35]. See also [M 1]	1	Acrylamide 30.0 g Bis 0.8 g	8.2 p. water 3.0 p. No. 1 0.8 p. No. 2 0.3 p. Ampholine	4.1 p. water 3.0 p. No. 1 0.8 p. No. 3 4.1 p. No. 4 0.3 p. Ampholine	0.2% sulfuric acid	0.4% ethanolamine	Per standard gel column (i.d. 5 mm, length 65 mm) up to 2 mA constant current with voltage increasing up to 350 V, 3 hrs. 20° C
	2	TEMED 1.0 ml Riboflavin 14.0 mg					
	3	TEMED 1.0 ml					
	4	Persulfate 70.0 mg					
No. 2 acc. to AWDEH [A 46], modified [G 32b]	5	Acrylamide 33.3 g Bis 0.9 g	30 p. water 6 p. No. 5 2 p. No. 6 2 p. Ampholine		5% (v/v) phosphoric acid	5% (v/v) ethylenediamine	Per gel plate of 170×260×1 mm constant voltage of 330 V for 5–7 hrs. with current decreasing from 30 to 2-3 mA
	6	Riboflavin 4.0 mg					
No. 3 acc. to CATSIMPOOLAS [C 13]	7	Acrylamide 25.0 g Bis 1.0 g	4.5 p. water 2.0 p. No. 7 0.5 p. No. 3 2.5 p. No. 6 0.5 p. Ampholine		5% phosphoric acid	5% ethylenediamine	Per standard gel column constant voltage of 150 V with current decreasing from 5 to 0.5 mA, 1 hr.
No. 4 acc. to ALLEN [A 22a]	8	Acrylamide 24.0 g Bis 0.84 g		6.85 p. water 10.0 p. No. 8 2.4 p. No. 3 0.75 p. Ampholine 10.0 p. No. 4	0.1 M HCl (cold)	0.15 M ethanolamine (cold)	Per gel plate of 750×100×3 mm, 100 V, 0.1 MFD, 90–95 pps with pulsed power supply, 6–10 hrs. for pH 3–10 gradient

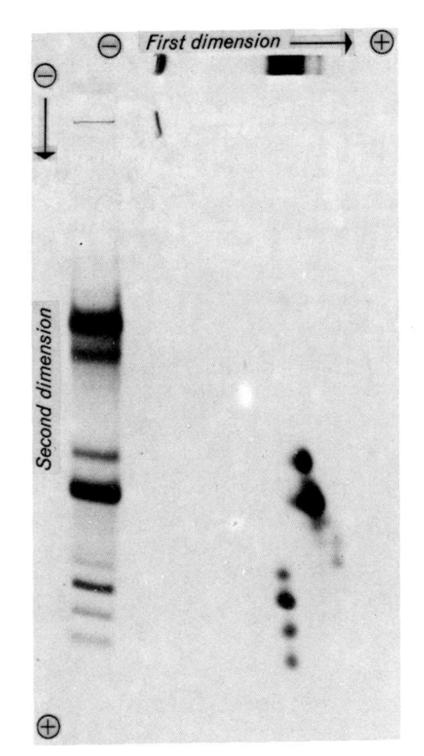

Fig. 59. Combination of isoelectric focusing and electrophoresis in polyacrylamide gel for two-dimensional analysis of potato tuber sap peroxidases *(protein-mapping)*. From: STEGEMANN and MACKO, see [M 1] for methods.

First dimension: Electrofocusing in a gel (T = 7.5 %, C = 5 %, i. d. = 5 mm) containing 2.5 % Ampholine 40 % pH 3–5, at 1 mA, 50–100 V for 4 hrs. Second dimension: Electrophoresis in a 3 mm thick gel slab (T = 5 %, C = 5 %) containing 0.125 M Tris-borate buffer pH 8.9. Dianisidine-H_2O_2-staining for peroxidases. The upper horizontal gel rod shows the isoelectric enzyme pattern after staining, the "empty" gel below served as starting line for electrophoresis. Left: electrophoretic pattern run simultaneously.

may be applied. If a sample solution is to be added, the prescribed amount of water of the working solution must be reduced correspondingly. In case of polymerization artifacts (see page 59), reasonable concentrated protein solutions (with 10 % sucrose) are put on top of the preformed gel under a protecting layer of Ampholine (2.5 % in 5 % sucrose). For this procedure [W 35] gels are prepared with protein omitted. About 10 mm of space left in the top of the tube are filled with the protecting layer. Current is passed for about 30 min. to set up the pH gradient. Finally the sample is carefully layered onto the gel under the Ampholine layer.

Samples to be applied on horizontal gel plates are simply pipetted onto the surface (100-400 μg protein in < 50 μl solution) and spread over a rectangular area [A 46]. The samples may also be soaked into paper pieces which are then laid onto the surface [G 32b]. For vertical gel plates using sample wells the position of the well bottom is critical [A 22a]: Since the well bottom will be at some pH gradient at the end of the run, a given protein may be poorly resolved if its IP coincides with the well bottom. Hence, the location of the well bottom should be raised or lowered accordingly. In this procedure, the sample is polymerized into working gel solution in the sample well after water layering. More well gel is then added up to the top. An optically flat bottom of both the gel and the sample well improves resolution.

5.2.2. Staining and Destaining Procedures

The electrofocused gel is stained, according to GRÄSSLIN [G 32b], by gentle agitating for 30-60 min. in a solution of 0.2% Coomassie Brilliant Blue G 250 (see page 74) in ethanol-water-glacial acetic acid (45:50:5, V/V) and destained with a mixture of ethanol-water-glacial acetic acid (40:55:5, V/V) during about 2 hrs. with 2 wash changes.

Another procedure [A 22a] involves fixation in 12.5% TCA at 65° C for 30 min., staining in a mixture of 0.2% aqueous Coomassie-ethanol-glacial acetic acid (45:45:10, V/V) at 65° C for 30 min. and destaining in ethanol-water-acetic acid (25:65:10, V/V) at 65° C for 20 min. Preservation in 10% acetic acid. Proteins may also be fixed by 20% sulfosalicylic acid (1 hr.) and stained by an 0.01% Coomassie solution in ethanol-water-acetic acid (7:11:2, V/V) overnight. Destaining and preservation preferably in ethanol-acetic acid-water-glycerol (5:2:13:5, V/V). Staining with 0.5% Amido Black in 7% acetic acid requires that the ampholytes are completely removed, prior to staining, by exhaustive washing with several changes of 12% TCA, since Amido Black stains the Ampholytes as well. — For a less sensitive method using bromophenol blue see [A 46a].

5.2.3. General Comments

To determine the pH range, gel sections of an unstained electrofocused gel are suspended in 1 ml water for 1 hr. [C 13]. The pH is measured with microelectrodes at 25° C.

If the molecular sieving effect of the gel inhibits migration of high molecular weight proteins, a 0.5% agarose-2% polyacrylamide matrix may be utilized as well [R 41] (see page 66). Methylred may be added to the cathode (lower) electrode bath as tracing dye to a final concentration of 0.5 ppm [M 1]; it forms a yellow band in the gel which turns red after reaching the upper gel part.

6. APPLICATIONS IN CLINICAL CHEMISTRY

6.1. Blood Proteins

6.1.1. Serum Proteins

6.1.1.1. General Remarks

The high separative power of disc electrophoresis becomes particularly evident in comparison with the well-known paper electrophoretic pattern of normal human serum proteins. While paper electrophoresis covers the so-called classical five serum proteins (albumin, α_1-, α_2-, β- and γ-globulins), disc electrophoresis separates up to 30 fractions [e.g. D 7, M 56, P 6, T 5]. Using discontinuous gradient gel electrophoresis (p. 62), WRIGHT and MALLMANN [W 36–37] increased this number up to 62.

Agar immunoelectrophoresis and disc electrophoresis were found to possess approximately equal resolving capacities [I 3]. However, since agar immunoelectrophoresis mostly furnishes only semiquantitative data and the efficiency of the precipitation reaction depends on the type and quality of the antiserum utilized, this method is by and large restricted to the investigation of certain diseases (e.g. paraproteinemias) and special problems of clinical research. The advantages of disc electrophoresis suggest that the method is also suited for *routine diagnostic analysis*. Although this is true in principle, the general applicability of disc electrophoresis for this purpose is hampered by difficulties mainly coming from the diverse possibilities for interpreting the complicated protein patterns. For the time being it is not yet possible for all cases to offer a satisfactory interpretation of the complex patterns.

An example of the *difficulties in interpretation* is the overlapping of α_2-globulins with β- and γ-globulins. [I 3] (Figs. 60 and 61). The γ-globulins represent a band complex which is difficult to evaluate since it extends as a "protein background" for α_2- and β-globulins from the start to the α_2-zone.

To overcome some of these difficulties a principle prerequisite may be that the electrophoretic separation conditions as well as recovery and treatment of the sample materials are carefully standardized. Standardization of gel formation is another important aspect (see page 55). Furthermore, it appears to be necessary to identify and characterize the numerous protein bands clearly and sufficiently. A new *nomenclature* might help to clarify this problem.

For this reason, *reference systems* are useful to facilitate the assignment of the numerous bands. DOWDING and TARNOKY [D 30, T 5] proposed a millimeter system with albumin as reference; for this purpose, they enlarge the photographs of their separation gels such that the distance between the start of the separation gel (0) and the forward albumin boundary (100) always amounts to 100 mm. PASTEWKA et al. [P 6] refer their values to the position of transferrin C (Fig. 62). FELGENHAUER et al. [F 14] take the symmetrical albumin peak as reference value 100. Anyhow, such normalized patterns are convenient for interlaboratory reference and are in digital form suitable for computer input.

Altogether with the subtle differentiation of protein patterns as offered by disc electrophoresis a refinement of the way of questioning in clinical research appears to become necessary. For example, in distinct cases statistically significant differences of protein patterns may be only obtained if the stage of a disease is taken into account. In addition, genetic controlled protein variants (e.g. haptoglobin and group specific components) which occur in normal serum should be considered. It is here where disc electrophoresis has its limits now. These problems may be solved, however, by routine cataloguing on a large scale. There are reasons to assume that the high resolving capacity of the method will yield a refinement of the routine laboratory diagnosis as expressed by a satisfactory specificity for a particular disease. For this aim extensive, statistically confirmed series of experiments and long term computerized memory banks of instantly recallable individual and group patterns will be required to improve the reliability and make use of the great informative value of disc electrophoresis for diagnosis and therapy control.

6.1.1.2. Normal Sera

For diagnostic purposes, it is often required to compare the patterns obtained by different methods, such as paper, cellulose acetate, agar gel, starch gel, polyacrylamide gel and agar immunoelectrophoresis. The question of the assignment of the different proteins therefore deserves a thorough experimental clarification. This is especially true for disc electrophoresis in view of the large number of fractions produced. FELGENHAUER [F 13a], among others, has identified several human serum proteins by special staining, enzymatic and immunological techniques (Fig. 60).

ISICHEI [I 3], using a preparative, free-flow electrophoresis apparatus according to HANNIG [H 7], isolated the 5 classical serum protein fractions and subjected them individually to disc electrophoresis (Fig. 61). With the exception of the α_1-fraction containing a relatively large amount of albumin, all proteins could be considered as homogeneous after cellulose acetate and agar immunoelectrophoresis; however, by disc electrophoresis the α_2-globulin fraction could be resolved into 10 and the β-globulin fraction into 7 subfractions. In contrast to paper, cellulose acetate and continuous free-flow electrophoresis, the α_2-bands exhibited lower migration velocities than the β-globulins due to the molecular sieving effect of the gel (page 4): Because of their molecular "bulkiness", α_2-globulins are slowed down in the gel network to a greater degree than β-globulins in spite of their higher charge.
BRACKENRIDGE and BACHELARD [B 53] found optimal resolution of human serum proteins in gels of T = 7 % and C = 1.71 % (6.88 % acrylamide and 0.12 % Bis).

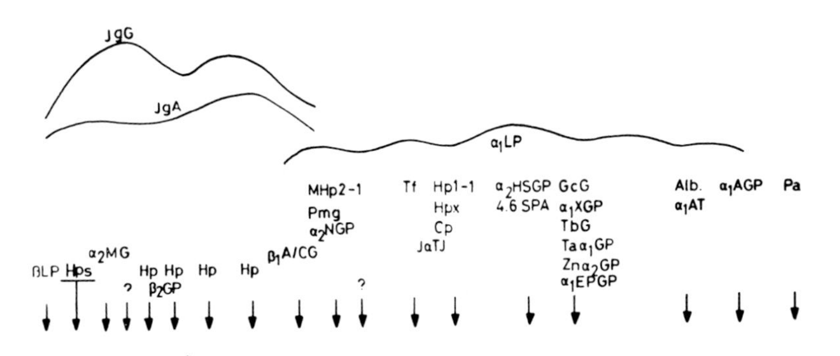

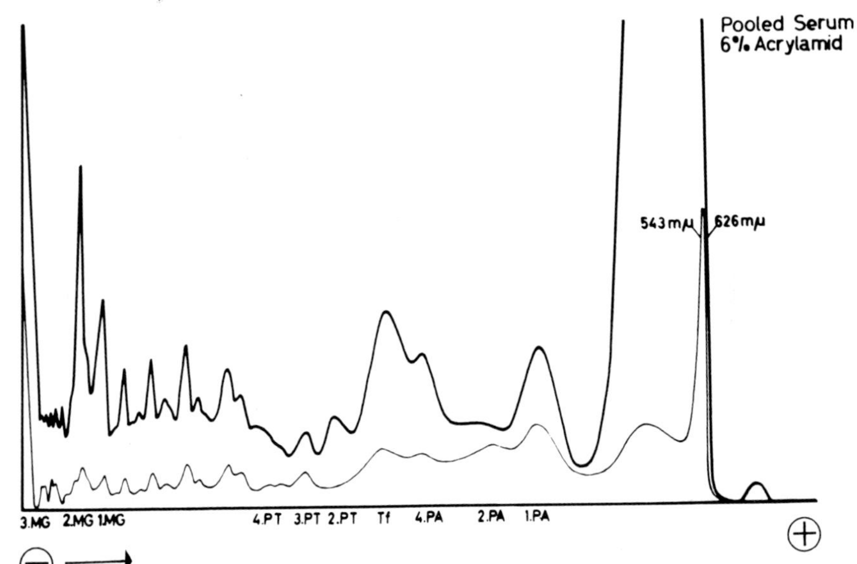

Fig. 60. – Position of several human serum proteins after disc electrophoresis on 6 % gels. Identification by special staining, enzymatic and immunological techniques. From: Felgenhauer [F 13a].

Pooled normal serum. Amido Black (626 nm) and PAS (543 nm) staining for protein and glycoprotein detection respectively; both densitometer graphs are optically synchronized to the same length.

Alb = Albumin
α_1AGP = Acid α_1-Glycoprotein
α_1AT = α_1-Antitrypsin
α_1EPGP = Easily precipitable α_1-Glycoprotein
α_1LP = α_1-Lipoprotein
α_1XGP = α_1-X-Glycoprotein
α_2HSGP = α_2-HS-Glycoprotein
α_2MG = α_2-Macroglobulin
α_2NGP = α_2-Neuramino-Glycoprotein
βLP = β-Lipoprotein
β_1A/CG = β_1-A/C-Globulin
β_2GP = β_2-Glycoprotein
Cp = Ceruloplasmin
GcG = Gc-Globulin
Hpx = Hemopexin
Hp(s) = Haptoglobin (s)
IgA = Immunoglobulin A
IgG = Immunoglobulin G
IαTI = Inter-α-trypsin inhibitor
MG = Macroglobulin
MHp 2–1 = Monomer haptoglobin 2–1
PA = Postalbumin
Pa = Prealbumin
4,6SPA = 4,6S Postalbumin
Pmg = Plasminogen
PT = Posttransferrin
Taα_1GP = Tryptophane-poor α_1-Glycoprotein
TbG = Thyroxine-binding globulin
Tf = Transferrin
Znα_2GP = Zn α_2-Glycoprotein

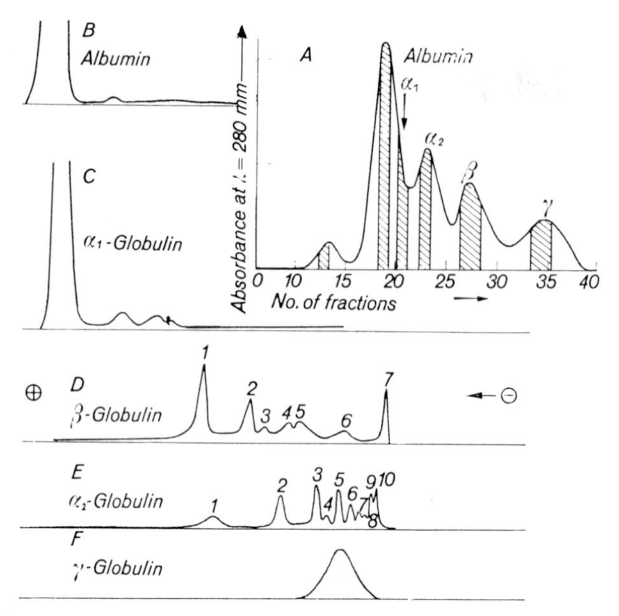

Fig. 61. – Disc electrophoretic fractionation and identification of serum proteins separated by preparative free-flow electrophoresis. From: ISICHEI [I 3].

(A) Elution profile of preparative electrophoresis acc. to HANNIG [H 7] (2 ml normal serum, Tris-citrate buffer, pH 8.6, 2300 V, 160 mA, +6° C); shaded = collected fraction. (B–F) Disc electrophoretic separation of collected and concentrated fractions. Densitometric evaluation of the gel photographs.

By means of gel system No. 1a (pp. 44-45), *prealbumin, albumin, postalbumin, transferrin and the multiple haptoglobins* can be easily identified in a normal serum pattern, while the *γ-globulins* form a broad zone from the start of the separation gel up to transferrin [for example: C 18, D 7, F 14, O 9, P 6] (Fig. 62). *β_1-lipoprotein* (Sβ) is found at the interface between spacer and separation gel, followed by *S-α_2-globulin* (19 S-α_2-glycoprotein) as the next main band (in the separation direction); *haptoglobins* of the 2–2 type are distributed inbetween. They are followed by 7 *S-γ-globulins*, other haptoglobins and globulins, which are not specifically characterized, up to transferrin. 19 *S-γ-globulins* are retained on the interface between sample and spacer gel. Free *hemoglobin A*, if present in the serum, migrates to the zone of the slow postalbumins. The lines of the *Gc types* are in the α_1-(postalbumin) region [T 5–6].

Alpha-1-glycoprotein, transferrin, α_2-macroglobulin, β_{1A}- and β_{1C}-globulins [G 7], β_2-glycoprotein [F 21], and β_{1C}-globulin [A 31] have been isolated and analyzed. Further studies were reported on transferrin [B 57, J 5], the renal action of anionic fractions of plasma globulins [L 34], hemopexin (β-glycoprotein) [H 10], a thyroxine-binding globulin [G 9] etc.

The normal serum protein pattern varies with age and genetically determined protein variants. According to TARNOKY and DOWDING [T 5–6], there is a progressive change in pattern throughout life. Similarly, MAN and WHITEHEAD [M 4] reported about serum proteins in infancy and pregnancy, MARGOLIS and KENRICK

[M 14] on the oestrogen dependent occurrence of a pregnancy protein in sera of pregnant women and women taking oestrogen containing pills.

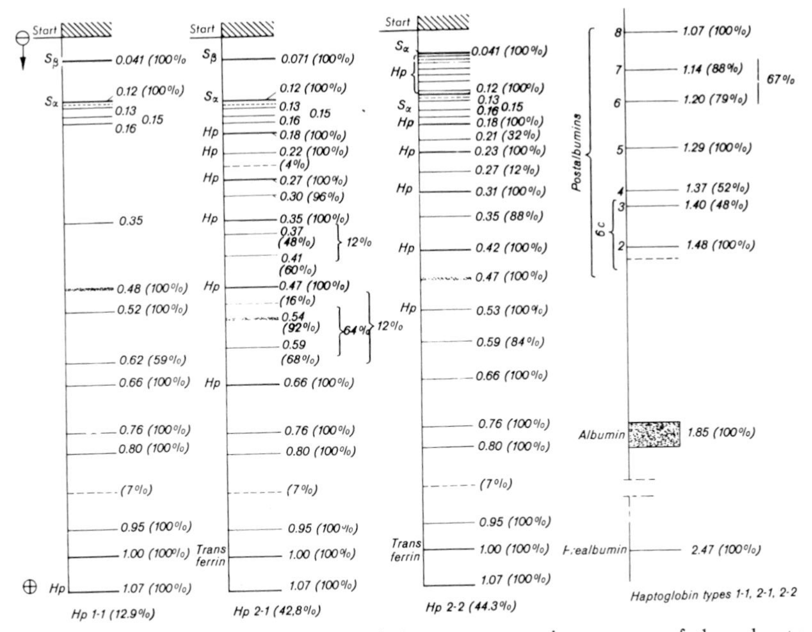

Fig. 62. – Graphic representation of the serum protein patterns of three haptoglobin types according to PASTEWKA et al. [P 6].

Normal adult sera. Relative migration distances referred to transferrin = 1.00. Frequency of occurrence in per cent (in parentheses). Dashed lines: rare bands. No data for relative protein concentrations. Hp = haptoglobin, Gc = group-specific component, Start = start of separation gel. Gel system No. 1a, Amido Black staining.

Using discontinuous gradient gel electrophoresis (see page 62) WRIGHT and MALLMANN [W 36] found 44–50 protein bands in normal serum.

For this purpose, tubes of 12 cm length are prepared with the following layers: 4.5 cm 10 % separation gel, overlayered with 3.5 cm 4.75 % separation gel (both with 0.4 % Bis), followed by 2.0 cm spacer gel; other conditions as for gel system No. 1. A three layer gel containing 16, 4.75 and 3.75 % acrylamide may yield up to 62 fractions [W 37].

Quantitative, statistical evaluations of normal protein patterns were reported, for example, by MOGI [M 56] and FELGENHAUER [F 14], using 52 and 55 sera, respectively. See also [L 1].

By disc electrophoresis PASTEWKA et al. [P 6] were able to assign 200 normal adult sera to their *haptoglobin type* with a reliability of 90 % and to establish the frequency and deviation of individual haptoglobins (Fig. 62). The reliability of identification referred to a standard method for haptoglobin classification. The significance of polyacrylamide gel electrophoresis for an improved Hp-type diagnosis was also demonstrated by several other investigators using continuous and discontinuous buffer systems [e.g. F 15, H 2, J 4, M 16, S 14, S 41, Z 5].

6.1.1.3. Pathological Sera

Serum protein electrophoresis probably found its most important application in the investigation of paraproteins. Consequently, the protein patterns in *dys-* and *paraproteinemia* (reaction configurations of WUHRMANN and WUNDERLY [W 18]) as acute infections, subacute-chronic-inflammatory and proliferative processes, neoplasms, hepatitis, cirrhosis of the liver, obstructive jaundice, nephrotic syndromes, β_1 and γ-globulin myeloma (γ_{1A}-myeloma and γ_{1M}-macroglobulinemia, Waldenström's disease) have aroused growing interest.

Three main components of the γ-globulin system can be distinguished: γ_2 (7 S)-, γ_{1A}(β_{2A}, 7–17 S)- and γ_{1M} (β_{2M}, 19 S)-globulins. The myeloma serum shows an tncreasing percentage of γ_2- and heterogeneous γ_{1A}-globulins. In the past, a distinction between myeloma proteins and Waldenström's macroglobulins has only been possible with expensive methods, such as ultracentrifugation, agar immunoelectrophoresis and starch gel electrophoresis. The studies of ZINGALE et al. [Z 6], CLARKE C 41], HAMMACK et al. [H 5] and KOCHWA et al. [K 25] have shown that disc electrophoresis proves to be a useful supplementary method for differential diagnosis (generally with gel system No. 1).

ROHLFING et al. [R 50] observed changes of distinct serum proteins stained with Coomassie in hematologic neoplasms. Similarly, AMARAL [A 26] found differences between the patterns of normal and chronic lymphatic leukemia sera. He also showed that disc electrophoresis gave all laboratory data needed to diagnose nephrotic syndrome in the presence of myeloma, while the cellulose acetate pattern was confusing. ZEIDMAN et al. [Z 2] reported on serum protein changes in neoplasia, DITTMAR et al. [D 22] on the coexistence of polycythemia vera and biclonal gammopathy with two Bence Jones proteins, and DORNER et al. [D 28] on H- and L-chains, purified human antibodies and myeloma proteins.

MOGI [M 57] investigated 180 pathological sera (Fig. 63) from diseases of the liver, such as obstructive jaundice, cholecystitis, acute hepatitis, Banti's syndrome, Wilson's disease, hemachromatosis; nephrotic syndrome; cancer of the stomach and peritonitis carcinomatosa; and inflammatory diseases, such as rheumatic fever and rheumatoid arthritis. ISICHEI [I 3] reported on lipoid nephrosis (Fig. 64). TARNOKY and DOWDING [T 5–6] analyzed 350 serum patterns of a great variety of diseases. Although the statistical basis of any one examined disease was too small to produce firm diagnostic patterns that would hold over a large population, it was the authors' experience that disc electrophoresis in clinical medicine may be particularly useful in Hp and Gc typing, classification of sarcoidosis and multiple myeloma, control of herapy in myelomatosis and hepatitis, and finding the primary site of metasizing carcinoma.

Plasma proteins were subjected to disc electrophoresis in cases of schizophrenia [S 3], nephrotic syndrome and cystic fibrosis of the pancreas [C 41], following preparative chromatography on DEAE-cellulose [O 3], after binding with chromium salts [P 19] and after proteolysis with bromelin [C 19]. In addition, basic lymphocyte proteins have been analyzed in a case of chronic lymphadenosis [W 12].

PRICE et al. [P 28] examined the *seromucoid patterns* of patients suffering from various chronic diseases, such as rheumatic fever with carditis, myocardial infarct, pneumococcal pneumonia, and schizophrenia. Statistically significant differences between normal and tubercular seromucoids were found by quantitative densitometry [P 29] (Fig. 65). The separations showed a satisfactory reproducibility: the coefficient of variation fluctuated between 6 and 20 %

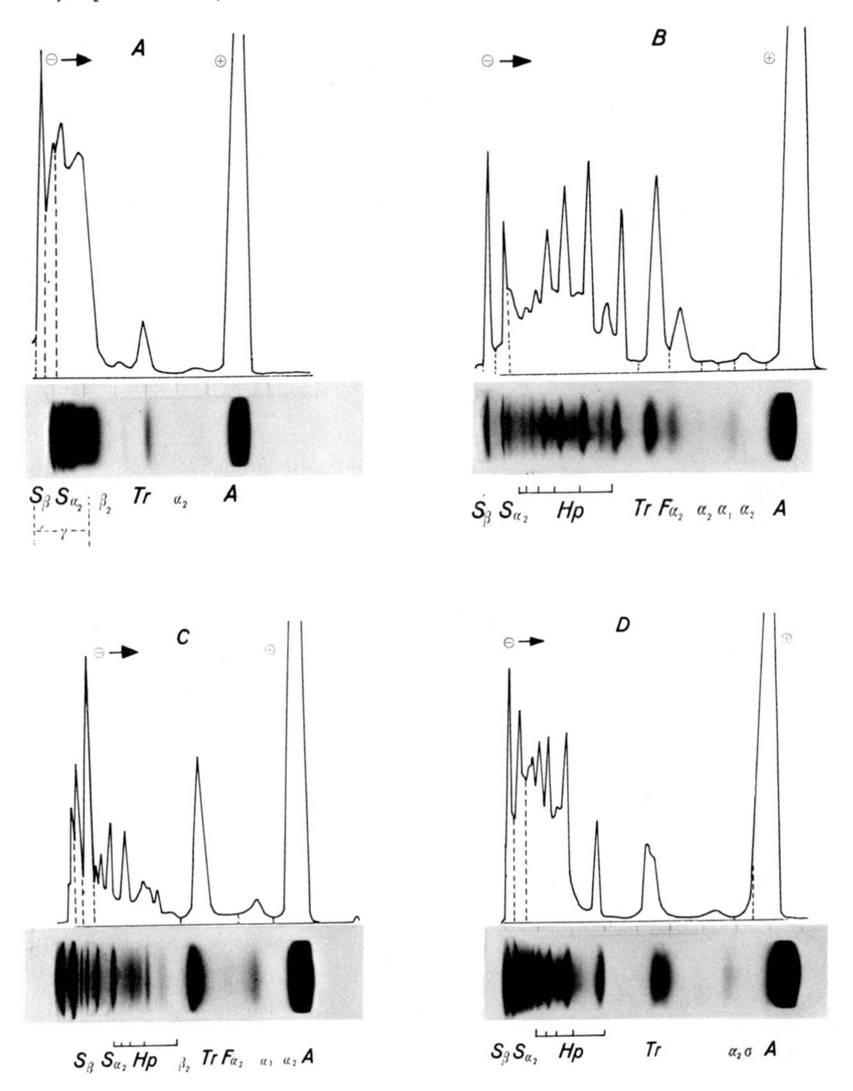

Fig. 63. – Disc electrophoretic patterns of pathological adult sera: (A) Cirrhosis of the liver; (B) obstructive jaundice; (C) nephrotic syndrome; (D) cancer of the stomach. From: MOGI [M 57].

3 μl serum, gel system No. 1, electrophoresis at 2.5 mA/gel, 135 V, 1.5 h. Amido Black staining.

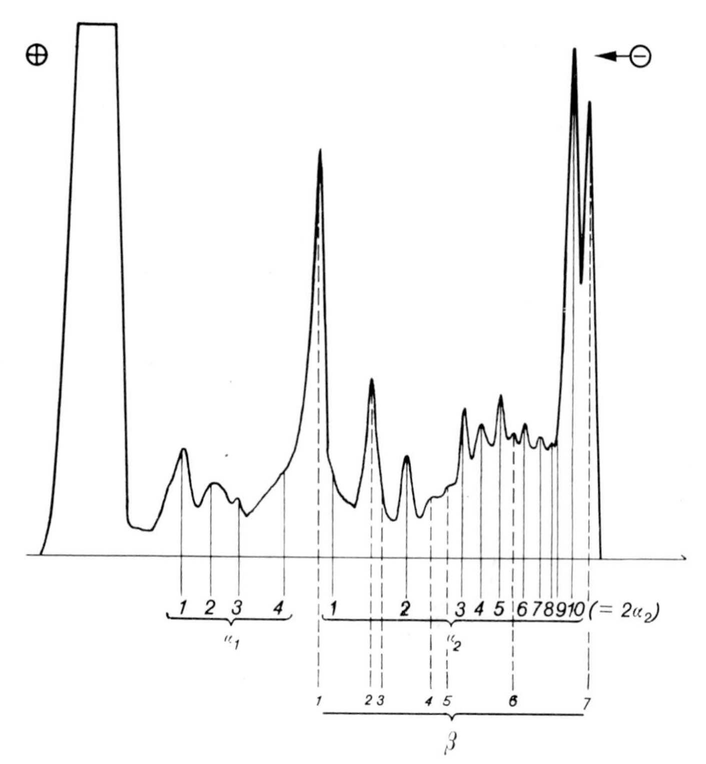

Fig. 64. – Disc electrophoretic pattern of a lipoid nephrosis serum with indication of the fractions. From: ISICHEI [I 3].

β_7-lipoprotein and all 10 α_2-bands are present in increased amounts. During therapy, all bands regress except for $\alpha_{2,10}$-globulin; the latter even increases and consequently maintains a constant total α_2-globulin content of the serum! – Gel system No. 1a, 4 mA/gel, 1.5 h. Amido Black staining. Densitometry of the gel photograph.

in a comparison of corresponding bands. KEYSER [K 14] separated seromucoids in a continuous polyacrylamide gel system, RENNERT [R 28] acid *mucopolysaccharides* (chondroitin and heparitin sulfate). Several reports present data about specific fraction patterns which may be used now or in the near future as clinical tests for diagnosis or prognosis: serum mucoproteins in malignant neoplastic diseases [B 1], beta-1-glycoprotein resembling orosomucoid from plasma, urine and leucocytes in rheumatoid arthritis and chronic myeloid leukemia [K 33], large glycoproteins in macroglobulinemia [M 35], and glycomucoproteins in pregnancy [A 5].

HILBORN and ANASTASSIADIS [H 20a] found that acidic mucopolysaccharides can be separated, at best, in highly alkaline 6 % gels (pH 11.5).

NARAYAN et al. separated *lipoproteins* [N 10, N 14, N 16] by disc electrophoresis in 3.75 % separation gels and *chylomicrons* [N 12] in spacer gels (with an average pore size of 200 mμ), while RAYMOND et al. [R 12] preferred continuous buffer systems to analyze lipoproteins in 3 %-gels. On lipoproteins from erythrocyte stroma see [M 65], on normal and pathological plasma lipoprotein patterns analyzed by polyacrylamide gradient gel electrophoresis [P 27]. See Table 7, page 50 for a suitable gel system.

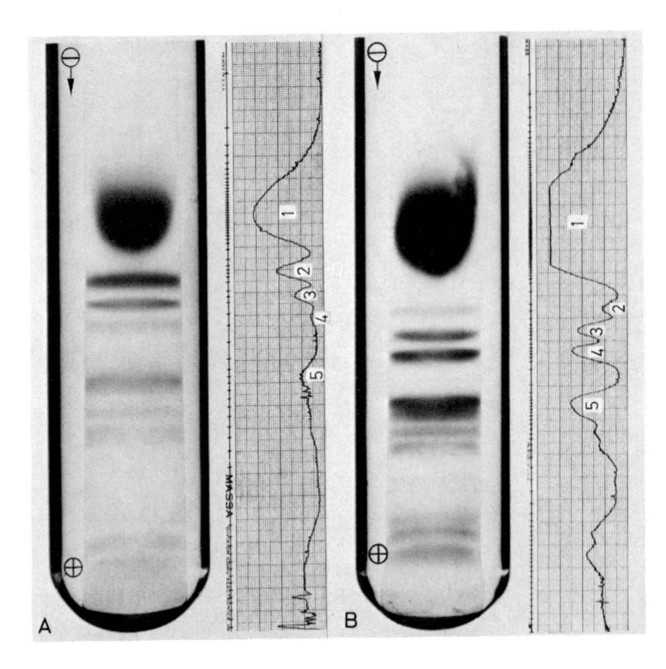

Fig. 65. – Seromucoid pattern of a normal serum (A) and a *tuberculosis* serum (B). From: Price et al. [P 29].

Bands 1–5 show statistically significant differences between the two sera (Gel system No. 1a). Electrophoresis at 3.75 mA/gel, 20° C, Amido Black staining.

Canalco (firm 5) offers a special gel reagent kit for *routine lipoproteinemia diagnosis*. It is designed for maximum productivity and permits rapid processing, thus making electrophoretic screening of large patient population economically feasible. The Canalco Quick-Disc (QDL) system provides a 30 min. test to identify the five different types of hyperlipoproteinemia, each of which requires different diet and drug therapy for successful treatment [L 21]. The test is an important element of the general *Atherosclerosis Potential (AP) test* (Fig. 66).

Similar *Quick Disc clinical reagents kits* are available from Canalco for rapid diagnosis of *hemoglobinopathy* from whole blood without preparation of hemolysate, separation of serum *lactic dehydrogenase isozymes*, screening of abnormal monoclonal or polyclonal *gammopathy*, *dilute proteins in urine* (Bence-Jones), *spinal fluid* (neurological disorders) and other sources without preconcentration and desalting required, *fine structure* study of specific subtypes related to genetic or disease conditions (haptoglobins in nephrosis, post-transferrin in chronic lymphatic leukemia, C-reactive and mucoproteins in cancer, γ_M-globulins in Waldenström's macroglobulinemia etc.).

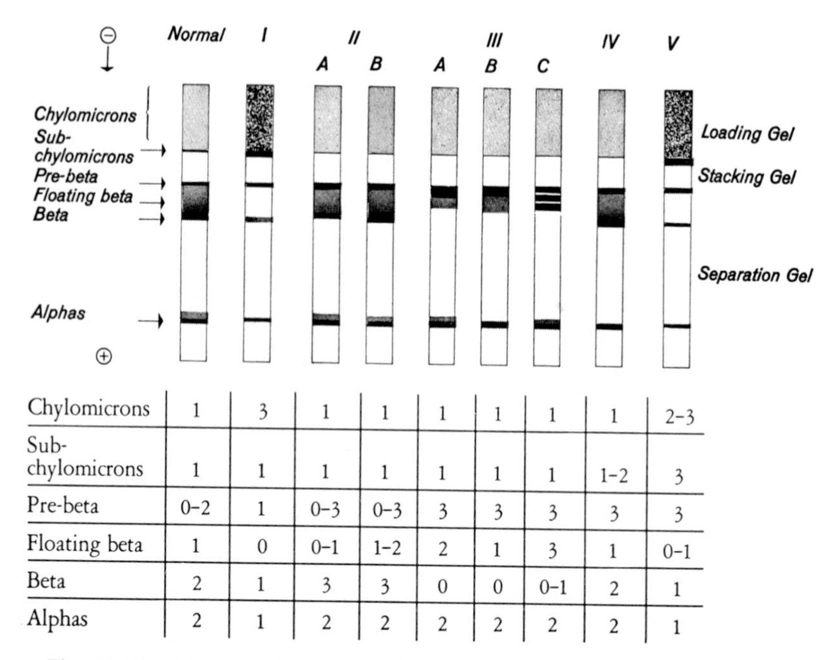

	Normal	I	II A	II B	III A	III B	III C	IV	V
Chylomicrons	1	3	1	1	1	1	1	1	2–3
Sub-chylomicrons	1	1	1	1	1	1	1	1–2	3
Pre-beta	0–2	1	0–3	0–3	3	3	3	3	3
Floating beta	1	0	0–1	1–2	2	1	3	1	0–1
Beta	2	1	3	3	0	0	0–1	2	1
Alphas	2	1	2	2	2	2	2	2	1

Fig. 66. Identification of normal and pathological lipoprotein patterns by disc electrophoresis for hyperlipoproteinemia diagnosis as part of the Atherosclerosis Potential (AP) test. Acc. to Canalco (firm 5) QDL instructions.
Band intensity: 0 none (absent), 1 weak, 2 moderate, 3 strong.

In summary, it can be said that disc electrophoresis is at least equivalent to agar immunoelectrophoresis with regard to the resolving capacity, but that a definitive evaluation of the method for purposes of differential diagnosis would be premature in view of the small number of investigated cases. The present findings justify large-scale quantitative statistical studies which would produce information on the applicability of the method for routine clinical diagnosis. The necessary prerequisites toward this end are discussed on page 137.

6.1.1.4. Immunoglobins, Antibodies and Complements

Hokama [H 33–34] and Riley [R 40] were able to identify and characterize *C-reactive* (*CRP*, *acute-phase*) *protein* by disc immunoelectrophoresis and found multiple molecular forms of the protein (Fig. 67). For other studies on CRP see [C 54] and [N 37]. Holland et al. [H 35] pointed out the significance of antibodies which occur in the form of precipitins of lactoproteins in the serum of children with respiratory infections, anemias and hepatosplenomegaly. The heterogeneity of the L- ($\varkappa$- and λ-) and H- (γ-)-polypeptide chains of human *G-myeloma proteins* was demonstrated by Terry et al. [T 7] using 4 % polyacrylamide gels in the presence of 8.5 M urea (gel system No. 11), while

STEVENSON and STRAUSS [S 70] reported on the specific dimerization of the L-chains of immunoglobulins. SGOURIS et al. [S 32] used the method to test the purity of γ_G-immunoglobulins. METZGER [M 45] showed that a Waldenström γ_M-paraprotein exhibits antibody activity to humanγ_G-immunoglobulin. HONG and GOOD [H 39] studied immunoglobulins in hypogammaglobulinemia. *Complement* components were successfully analyzed by disc electrophoresis [L 2, N 38, P 17, P 23].

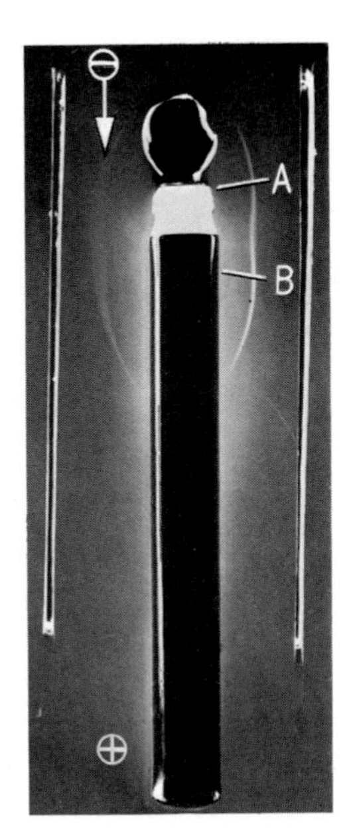

Fig. 67. Disc immunoelectropherogram of biphasic C-reactive protein (CRP). From: HOKAMA (unpublished).

Disc electrophoresis of a CRP-containing serum in gel system No. 1. Immuno diffusion with CRP-antiserum. Band A at the interface between sample and spacer gel; band B at the cathodic end of the separation gel.

6.1.1.5. Serum Enzymes

6.1.1.5.1. General Remarks

Studies in clinical chemistry using polyacrylamide gel electrophoresis have been carried out, for example, on esterases [H 47], cholinesterase in analbuminemia [D 17], cholinesterase as a genetic variant [N 24], cholinesterase in glaucoma serum [J 19], carbonic anhydrase of erythrocytes in newborns and adults [T 4], lactate dehydrogenase in respiratory acidosis [J 1], myocardial infarct, pulmonary thrombosis, liver diseases, and others [C 38, G 22], lactate and malate dehydrogenases in normal and leukemic leukocytes [R 2], 17β-hydroxysteroid dehydrogenase from placenta [J 3], phosphatases in neoplasms, liver cirrhosis, hepatitis [E 6], acid phosphatase in Gaucher's disease, carcinoma of the prostate and myeloma [G 13], and in lysates of blood cells, and homogenates of liver, spleen, lymph nodes and prostate [R 54]. Moreover three alkaline phosphatase isozymes were identified as liver, kidney and bone types [A 10].

During the separation of the isoenzymes of *lactate dehydrogenase* in 7% gels (gel system No. 1a), isozyme No. 5 cannot migrate in the pH range of the sample and spacer gel. (In endosmotic media, such as starch or agar gel, it migrates to the cathode.) However, if the sample and spacer gel are dispensed with and the enzyme solution (serum) is first treated with 40 % saccharose and then layered under the electrode buffer solution at 5° C (pH 8.3) on top of a 5.5 % separation gel (pH 8.9), readily reproducible separations are obtained as reported by DIETZ and LUBRANO [D 16]. They can be subjected to statistical analysis by quantitative densitometry if the migration distances are kept constant. Electrophoresis is run at 2.5 mA/gel for 45 min. with NBT staining taking 45–120 min. to obtain full color reaction.

6.1.1.5.2. Reactions for the Detection of some Enzymes

1. *Carbonic anhydrase*: According to KURATA [K 39], incubation proceeds for 30–60 min. at 37° C in a freshly prepared solution of 1 g manganese chloride ($MnCl_2 \cdot 4 H_2O$) in 100 ml of an 8 % aqueous sodium bicarbonate solution. (The manganese chloride solution is stirred until the pink color fades and is then filtered.) Subsequently, the gel is rinsed and incubated at 37° C in an 0.1 % aqueous potassium periodate solution until bands appear (several hours). TAPPAN et al. [T 4] prefer the method of PHILPOT and PHILPOT [P 18] (incubation at 25° C); for this purpose, they elute individual gel fractions.

2. *Cholinesterase*: According to CHRISTOFF et al. [C 37] electrophoresis is followed by washing the gel twice for 15 min. each by incubating it in 0.1 M Tris-maleate buffer pH 6.0 to remove separation gel buffer and phosphate ions. For staining, the gel is placed into an 0.03 % solution of diazotized p-chloroaniline in 0.1 M Tris-malate buffer of pH 6.0 for 30 min., followed by addition of the substrate (e.g. α-naphthylacetate; incubation concentration of 0.1 ppm = 0.01 mg %). Incubation proceeds for up to 12 hrs. at 4° C; the staining solution is replaced after 6 hrs. – DAVIS [D 9] also recommends a dithiooxamide (rubeanic acid)-copper method as described by PEARSE [P 10, p. 940], but in a modification developed for polyacrylamide gels: After careful washing with water, incubation proceeds first in saturated dithiooxamide solution in 50 % ethanol for 30 min.,followed by standing overnight in the saturated reagent solution in 0.02 M sodium acetate. The gel is washed with water in which the stain remains stable (see also [M 33]).

3. *Esterases* (*carboxyl esterases*): According to DAVIS [D 6], electrophoresis is followed by washing the gels twice for 10–15 min. each with 0.1 M Tris-HCl buffer of pH 7.0. For the preparation of the reagent solution, 2 drops of p-chloroaniline solution (36 mg/ml 1 N HCl) are mixed with 2 drops of sodium nitrite solution (26 mg/ml water) and 1 ml water (0° C) in an ice bath and the mixture is shaken well for 30 sec. This mixture is then added to 25 ml 0.1 M Tris-HCl buffer at pH 7.0 and 0.2 ml of a 1 % solution of α-naphthylacetate in acetone. The gels are incubated in the reagent solution with shaking (5 min. to 2 hrs.) until bands appear. – MARKERT and HUNTER [M 17] incubate the gels for 15 min. at 37° C in 5 ml substrate-stain mixture containing α-naphthylacetate, butyrate or propionate (40 mg/100 ml buffer) and Fast Blue RR salt (C.I. 37, 155; 70 mg/100 ml buffer) as a coupling reagent. The gel is then incubated in a mixture of 3 parts ethanol and 2 parts 10 % acetic acid for at least 30 min. at 37° C and stored in 10 % acetic acid (see also [H 46]).

4. *Catalase*: THORUP et al. [T 12] first dip the gel into a mixture of 28 ml 3 % hydrogen peroxide solution for 15–30 sec., after which it is immersed in 50 ml

1.5 % potassium iodide solution until all areas without catalase activity have been stained. The gel is rinsed with double-distilled water. It is important to prepare the buffer solutions with water which was redistilled in quartz equipment.

5. *Lactate and malate dehydrogenase* (*LDH and MDH*): ALLEN [A 14] incubates in darkness at 25° C in a mixture of 7.5 ml 0.05 M Tris-HCl buffer of pH 7.5, 3.0 ml 0.5 M DL-sodium lactate of pH 7.5, 1.25 ml 0.06 M KCN, 3.5 ml nitroblue tetrazolium (2 mg/ml buffer), 10 mg nicotinamideadenine dinucleotide (NAD^+) and 0.15 ml N-methylphenazonium methosulfate (2 mg/ml buffer). – According to GOLDBERG [G 14, G 19], the gels are washed with cold 0.1 M Tris-HCl buffer of pH 8.3 and incubated for 30–60 min. at 37° C in a mixture of 0.036 M L (+)-sodium lactate, 0.3 mg/ml NAD, 0.8 mg/ml nitroblue tetrazolium and 0.14 mg/ml N-methylphenazonium methosulfate in 0.1 M Tris-HCl buffer of pH 8.3. – SCHRAUWEN [S 23] rinses the gel with water and incubates it at 37° C for 15-60 min. in a mixture of 7.2 ml solution I, 0.2 ml solution II and 8 mg NAD pf pH 7.6. The mixture must be prepared daily from the solutions. Solution I: 1 ml sodium lactate (70–72 %) + 50 mg NaCN + 1.8690 g $Na_2HPO_4 \cdot H_2O$ + 0.2722 g KH_2PO_4 + 25 mg nitroblue tetrazolium (Nitro-BT); these substances are dissolved in distilled water; the solution is brought to 90 ml with distilled water and is filtered. Solution II: 10 mg N-methylphenazonium methosulfate (PMS) are dissolved in 10 ml distilled water. The two solutions I and II are stable for at least 14 days provided they are stored in complete darkness. The color development on LDH is visible already after about 15 min. Incubation at 37° C for more than 2 hrs. is not desirable. The same color development can be obtained by standing overnight at room temperature provided incubation takes place in darkness. After staining, the gel is rinsed with distilled water and stored in 2 % acetic acid. The color intensity of the bands does not change in a week. For the detection of malic dehydrogenase, 1 ml sodium lactate in solution I is replaced by a solution of 0.580 mg malic acid in 5 ml distilled water, neutralized with 0.1 N NaOH.

LAYCOCK et al. [L 8] found that the following incubating mixture was optimal for the detection of MDH: 5 mg/ml DL-malic acid, 1 mg/ml NAD, 0.1 mg/ml N-methylphenazonium methosulfate, 1 mg/ml MTT-tetrazolium (3[4,5-dimethylthiazolyl-2]-2,5-diphenyltetrazolium bromide) in 0.1 M phosphate buffer of pH 6.5.

6. *Leucinaminopeptidase*: PEARSE [P 10] as well as NACHLAS et al. [N 1] mix 3 ml substrate stock solution with 30 ml 0.1 M acetate buffer of pH 6.5, 24 ml 0.14 M NaCl, 3 ml 0.02 M KCN and 30 mg Fast Blue B-salt (tetrazotized o-dianisidine). The substrate stock solution is prepared by dissolving 24 mg L-leucyl-ß-naphtylamide (I) or L-leucyl-4-methoxy-ß-naphthylamide (II) first in a few drops of methanol, after which the volume is brought to 3 ml solution with water. The gels are incubated for 15 min. to 2 hrs. at 37° C, are rinsed for 2 min. in 0.14 M NaCl, immersed in 0.1 M copper sulfate solution for 2 min., and rinsed again with 0.14 M NaCl. Substrate I produces purple-blue and substrate II strong red bands.

7. *Lipase*: ABE et al. [A 2] treat a mixture of 5 ml 0.4 M Tris buffer of pH 7.4, 1.0 ml 2.5 % sodium taurocholate solution and 3.9 ml water with 0.1 ml of a 2 % solution of naphthol-AS-nonanoate (pelargonate) in dimethylacetamide and 10 mg Fast Blue BB salt (C. I. 37, 175) with vigorous stirring. The turbid solution is filtered. The gels are incubated for 40–60 min. at 37° C and washed with distilled water. (See also [M 69].)

8. *Peroxidase*: SCHRAUWEN [S 23] incubates the water-rinsed gel at 20° C for 15–45 min. in the absence of daylight in a mixture of 1 ml 0.2 M sodium acetate, 0.25 ml benzidine-guaiacol solution, 0.1 ml 5 mM manganese sulfate and 0.1 ml 0.12 % hydrogen peroxide. Benzidine-guaiacol solution: 50 mg benzidine + 135 mg guaiacol in 25 ml 10 % acetic acid, analytical grade. After about 10 min., the peroxidase containing zones on the gel surface assume a reddish-brown color; the maximum color depth is attained after about 30 min. The gel is then rinsed with distilled water and stored in 2 % acetic acid. If the gel is stored in the absence of light, the color intensity of the bands changes only little in a period of one week.

9. *Alkaline phosphatase*: By the *lead conversion method* [A 17], the gel is first incubated for 15 min. at 25° C in a mixture of 50 mM sodium-ß-glycerophosphate, 15 mM $CaCl_2$ and 33 mM Tris-HCl buffer at pH 9.5. It is then rinsed with water and incubated for 30 min. at 24° C in a mixture of 80 mM Tris-malate buffer of pH 7.0 and 3 mM lead nitrate. Incubation is followed by washing repeatedly with water for 1 hr. and then incubating for 2 min. in 5% ammonium sulfide solution. Finally, repeated rinsing with water takes place. – In the *azo-coupling method* [A 17], the gel is incubated for 15 min. at 25° C in a mixture of 25 mM sodium -α-naphthylphosphate, 1 mg/ml Fast Red TR (diazonium salt of 4-chloro-o-toluidine) and 33 mM Tris-HCl buffer at pH 9.5, followed by repeated washing with water to remove excess of substrate. – EPSTEIN et al. [E 6] report an indigogenic reaction stain procedure which is selective, sensitive, and quantitative (satisfactory linearity of staining). Electrophoresis lasts 30 min. at 2.5 mA/gel, staining overnight developing blue-green disks against a clear background. The advantages claimed are speed, sharpness of bands and elimination of the coupling step needed by other methods.

10. *Acid phosphatase*: ANDERSON et al. [A 29] recommend sodium-α-naphthylphosphate as a substrate and Basic Fuchsin as a coupling reagent. Since some acid phosphatases are thermally labile, the electrophoresis should be performed at 4° C (e.g. 2 mA/gel). Subsequently, the gels are washed twice for 15 min. each at 5° C with 0.1 M acetate buffer at pH 5.0. To prepare the incubating solution, 0.8 ml 4% Basic Fuchsin solution in 2 N HCl is mixed with 0.8 ml 4 % aqueous sodium nitrite solution; this mixture is added to 18 ml of a solution of 20 mg sodium-α-naphthylphosphate in 13 ml water and 5 ml veronal-acetate buffer (9.7 g sodium acetate · 3 H_2O + 14.7 g sodium diethylbarbiturate per 500 ml aqueous solution) and the pH is adjusted to 5.0 with 1 N NaOH. The solution is filtered. The gels are incubated in this solution at 4–5° C for 12–18 hrs. BARKA [B 6] incubates the gels for 12–16 hrs. at + 4° C in a mixture of 1 mg/ml sodium-α-naphthylphosphate, 1 mg/ml Fast Garnet GBC salt and 50 mM veronal-acetate buffer at pH 5.0. Barka found that the coupling reagents may differ considerably depending on their source; the best results are obtained with a product from firm 13. (See also [B 8].) Azo dye gives minimum background with some acid phosphatases when the enzymes are prefixed before staining in 10% neutral formalin acetate for 30 min. at 4° C [G 13].

6.1.2. Blood-Clotting and Platelet Proteins

Disc immunoelectrophoresis was used to analyze purified human plasminogen (profibrinolysin) [A 13]. See also [A 3]. Moreover, disc electrophoresis proved to be a valuable method to study a fibrinolytic digest fraction from fibrinogen [J 2], fibrinogen derivatives in pathological plasma proteolysis [F 23], platelet proteins [G 3], also in patients with Glanzmann's thrombasthenia [W 11], platelet proteases [N 2] and polymorphic prothrombin L [3–4].

6.1.3. Hemoglobin Determination

Disc electrophoresis permits the separation and quantitation of the *hemoglobin (Hb) A, F, S, A_2 and C* in 1 hr. without staining.

The hemolysate may be easily prepared from whole blood just before electrophoresis. According to the Canalco (firm 5) QDH instructions, the blood sample is mixed, on the top of the separation gel, with loading gel solution (sample and spacer gel solution) providing the concentrating effect and containing, in addition, a hemolysing agent (detergent). Thus, the blood is hemolysed just before photopolymerization of the sample gel. For another method of hemolysate preparation see [S 50]. The Hb fractions migrate to known relative positions within the separation gel (gel system No. 1a) and can be identified by their migration distance and, in some cases, by the quantity of Hb present in each fraction. Only the discontinuous buffer system of disc electrophoresis provides sharp Hb-A_2 separation from Hb-A as well as accurate A – A_2 quantitation for thalassemia diagnosis. The separation gels can be made up five days in advance of use without loss of resolution; in this case, they are immersed after polymerization, in stock solution No. 1 of gel system No. 1a. Other hemoglobinopathies such as Hb-AS or Hb-AC may be also identified by this technique combined with quantitative densitometry. For more detailed informations (including fetal and total Hb determination) see the QDH instructions of Canalco, for quantitative methods see also [N 21] and [S 9].

The modification of DOWDING and TARNOKY [D 31] which yields high resolution changes the order of mobilities of the hemoglobins from $C<A_2<S<F<A$ of gel system No. 1a to $A_2<C<S<F<A$. Instead of the Tris-glycine electrode buffer pH 8.3 the following solutions are used: Cathodic (upper) buffer: 25.2 g Tris, 2.5 g EDTA, 1.9 g boric acid in 1 l water, pH 9.1; anodic buffer: 5.15 g sodium diethylbarbiturate, 0.92 g diethylbarbituric acid in 1 l water, pH 8.6.

6.1.4. Ceruloplasmin

According to CLARKE [C 41], ceruloplasmin migrates in the β-globulin zone. SWEENEY [S 85] analyzed tissue and serum ceruloplasmin concentrations associated with inflammation, HOLTZMAN et al. [H 38] with WILSON's disease. For its detection see [O 13] and [M 64].

6.1.5. Nonhemoglobin Proteins of Erythrocytes

CAWLEY et al. found up to 18 nonhemoglobin proteins in erythrocytes. These proteins consist of several enzymes [C 19] and blood group substances [C 20].

SCHNEIDERMAN [S 15] solubilized proteins and lipoproteins from human erythrocyte stroma by treating the latter with Triton X-100. For further studies on erythrocyte membrane proteins see [L 5, R 53, S 16].

6.2. Proteins of other Body Fluids

6.2.1. Cerebrospinal Fluid (CSF)

The concentrating effect of disc electrophoresis (page 22) is of particular advantage for the analysis of CSF. With other analytical methods (such as continuous electrophoresis methods, ultracentrifugation) CSF needs to be concentrated first (for example, by high-pressure dialysis [B 63], acetone precipitation, ultrafiltration, lyophilisation, concentration under reduced pressure, etc.). This is due to the fact that normal blood serum contains 7–8 g, but normal CSF only 15–40 mg protein/100 ml fluid. That such concentrating procedure can actually produce artifacts in the protein pattern (presumably by denaturation or aggregation) was demonstrated by CUNNINGHAM [C 55]. Usually, 0.4–0.7 ml fluid is sufficient, although ASAGAMI et al. [A 42] consider 0.5 ml an optimal quantity. However, even 2 ml have been used per standard gel [W 9]. In most cases, a spacer gel of greater length is required in that case and, if such is utilized, also a longer sample gel (each about 5 cm in length).

CUNNINGHAM [C 55] and MONSEU and CUMINGS [M 61] use tubes of 11 cm length and prepare separation gels of 6 cm and spacer gels of 1 cm length; they add 25 % saccharose to 0.1–0.8 ml fluid and electrophorese at 2–3 mA/gel.

Several authors have described the normal and pathological CSF protein pattern [for example: C 50, D 12, E 8, F 14]. ASAGAMI et al. [A 39–42] and NAKAMURA et al. [N 5–7] analyzed patterns in meningitis, metastatic brain tumors, progressive paralysis, schizophrenia, etc. EVANS and QUICK [E 7] found deviation of the γ/β-globulin ratio in cases of infectious meningitis, chronic vascular diseases and tumor metastases; qualitative changes in cases of hemorrhage, serum exudation, and hydrocephalus. PAPADOPOULOS and SUTER [P 5] reported that disc electrophoresis furnishes useful supplementary information for the differential diagnosis of neuropathies (encephalitis, meningitis, Guillain-Barré syndrome, etc.). MONSEU and CUMINGS [M 61] found significant deviations from the normal pattern only in polyneuritis, but not in multiple sclerosis, whereas SHAPIRO et al. [S 36] observed the reduction or absence of a specific protein frequently occurring in spinal fluids from multiple sclerosis patients. GREENHOUSE and SPECK [G 34] found elevations of the β-globulins in acute and chronic polyneuritis, neoplasms of the central nervous system and atrophy of the brain as well as of the γ-globulins in destructive myelinopathies, acute bacterial and aseptic meningoencephalitis and Guillain-Barré's syndrome, but did not observe changes in the total protein content. The relationships between serum and CSF proteins in γ_2-and γ_{1A}-myeloma as well as macroglobulinemia (Waldenström) were investigated by WEISS et al. [W 10] and SCHEURLEN et al. [S 12]. The former authors found that γ_2-paraproteins can enter the CSF, while proteins of higher molecular weight ($>1.6 \times 10^5$) of the type of γ_{1A}-myeloma globulins and macroglobulins do so only with a damaged central nervous system (see Fig. 68). WEISS et al. [W 9] also investigated the influence of the CSF paraproteins on the colloidal gold reaction.

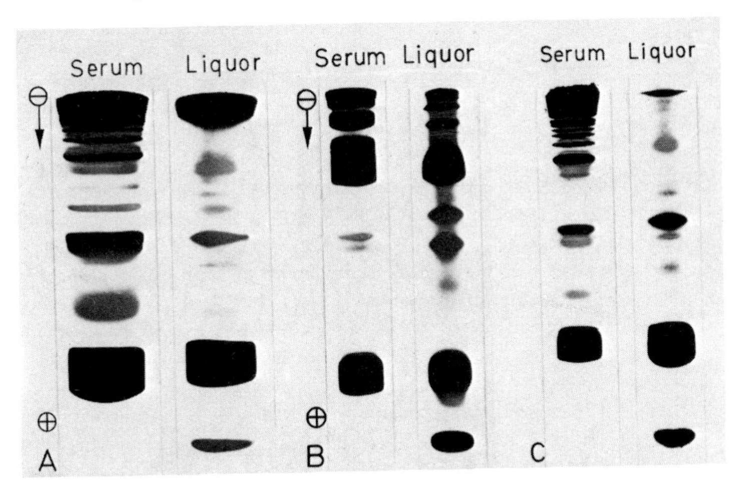

Fig. 68. – Myeloma and macroglobulinemia paraproteins in the serum and cerebrospinal fluid (CSF). From: WEISS et al. [W 10].

(A) γ_2-myeloma, paraprotein primarily in the "slow" γ-zone. (B) γ_{1A}-myeloma, paraprotein in the CSF (6 S). (C) Macroglobulinemia (Waldenström), absence of paraproteins in the CSF. Gel system No. 1, unconcentrated, sucrose-treated CSF (up to 2 ml) applied directly on the spacer gel. Amido Black staining.

6.2.2. Urine

A disc electrophoretic analysis of the urinary protein pattern permits a control of the effect of cytostatics on myeloma, for example, by the disappearance of *Bence-Jones proteins*. RIVERS [R 45] used 40 µl urine which were polymerized into a sample gel without concentration and without added sucrose. Generally, however, urinary proteins need to be concentrated also for disc electrophoresis, for example, by ammonium sulfate precipitation [K 19], concentration by dialysis, ultrafiltration, lyophilisation or benzoic acid adsorption. ADAMS-MAYNE and JIRGENSONS [A 6] used the first method to extract Bence-Jones proteins which they then investigated with gel system No. 7. See also [B 2]. LYTLE and HASKELL [L 42] adopted the benzoic acid method [K 9] to isolate a protein fraction containing *chorionic gonadotropin* from the urine of pregnant women; this fraction could be separated and purified by disc electrophoresis (Fig. 69). The isolation of the hormone is the basis for a pregnancy test.

CAWLEY et al. [C 21] and HAMASHIGE et al. [H 3] analyzed *chorionic gonadotropin* by disc immunoelectrophoresis. LESUK et al. [L 19] employed disc electrophoresis as a

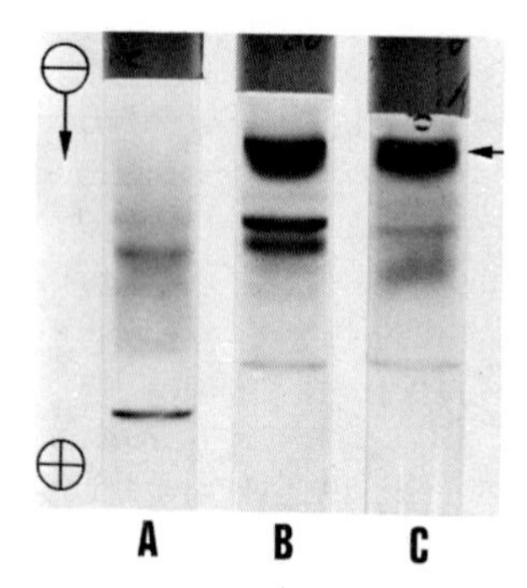

Fig. 69. – Separation of urinary proteins containing chorionic gonadotropin by disc electrophoresis. From: LYTLE and HASKELL [L 42].

(A) Control urine at midcycle; (B) urine from the same woman on the 52nd day, (C) on the 61st day of pregnancy. The arrow indicates the principal biological activity of the hormone.

purity test (gel system No. 8) in the isolation of *urokinase*, a plasminogen-activating urinary enzyme; with respect to resolving power, the method was superior to paper, celluloseacetate and starch gel electrophoresis. DUFOUR et al. [D 38] used the method to identify polymorphic *urinary phosphatases*, BOWDEN and CONNELLY [B 49] to analyse urine a-ketoaciddehydrogenase in ketoaciduria, and KAO et al. [K 5] to characterize urinary glycoproteins. ZINGALE et al. [Z 4] investigated serum and urinary proteins in various infantile kidney diseases, e.g. nephrosis, glomerulonephritis and uremic-hemolytic syndrome. KEUTEL [K 12] found polyacrylamide gels to be preferable to agar gels, when he compared urinary colloids by immunodiffusion. GRÄSSLIN et al. [G 32a] quantitated glycoproteins, plasma proteins and glycosaminoglycans of normal urine.

6.2.3. Saliva

CALDWELL and PIGMAN [C 4] and MANDEL [M 6] identified up to 21 different proteins in *parotid and submaxillary saliva* by disc electrophoresis. Moreover, STEINER and KELLER [S 67] reported on parotid proteins, DABBOUS and DRAUS [D 1] on submaxillary mucins. For analysis of human salivary amylase see [B 41]. According to MEYER and LAMBERTS [M 47], the method is particularly favorable because, unlike starch, polyacrylamide cannot be attacked by amylases. Thus, it was possible to separate blood group substance A, glycoproteins [C 3] and mucoproteins [A 34] from other salivary proteins. CALDWELL and PIGMAN [C 4] used gel system No. 1 and 20–30 μl saliva concentrated to 1/20 of its volume by lyophilization and dialysis. MOGI [M 58] analyzed salivary proteins in acute parotitis (gel system No. 1, 15–20 mm spacer gel for 0.15 to 0.2 ml saliva; electrophoresis at 2–3 mA/gel, 2–2.5 hrs.). MANDEL et al. reported on salivary proteins in cystic fibrosis [M 7a] and the effect of drugs on salivary secretion and composition [M 7].

6.2.4. Exudates, Transudates and other Body Fluids

Gel system No. 1 served for an investigation of fluids in pathological *synovial*, *ovarian*, *follicular*, *dental cysts*, and of *ascites* in carcinomatous peritonitis, thorax exudate in tuberculous pleuritis, effusions in the middle ear in otitis media and idiopathic hematotympanum, and nasal discharge during chronic sinus infection [M 58–59, N 7, O 2]. Different proteins in pancreatic juice were identified by KELLER and ALLAN [K 11] and *perilymph* proteins by PALVA and RAUNIO [P 1–2] on microgels. Similarly, *aqueous humour* proteins were fractionated by HEMMINGSEN et al. [H 18] on microgels. *Tear* proteins were analyzed by BONAVIDA et al. [B 42], *sweat* proteins by JIRKA [J 6], *seminal plasma* proteins by QUINLIVAN [Q 2], and other *genital fluids* by DAVAJAN and KUNITAKE [D 4]. WALES et al. [W 2] identified serum proteins in normal human *gallbladder bile* by disc immunoelectrophoresis.

6.3. Tissue Enzymes

The influence of *anticholinesterase compounds* on the zymograms of extracts isolated from tissues with a pathological end-plate (myasthenia gravis) was directly visualized and evaluated by CHRISTOFF et al. [C 38] by disc electrophoresis. Fig. 70 shows that the inhibitors 62 C 47 (10^{-5}M) and iso-OMPA (10^{-4}M) highly inhibit the activity of acetylcholinesterase and (pseudo)-cholinesterase. A *pharmacological-analytical method* can be based on this finding.

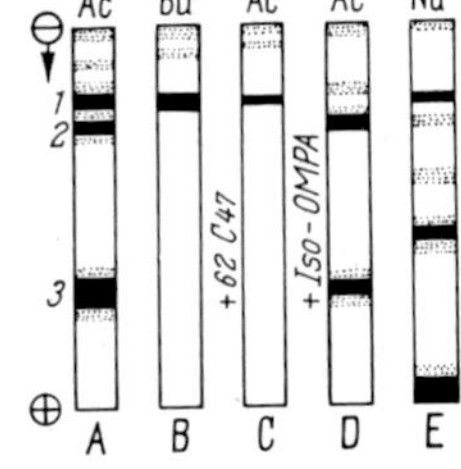

Fig. 70. – Influence of the cholinesterase inhibitors 62 C 47 and iso-OMPA on *cholinesterase zymograms* of rat gastrocnemic extracts. From: CHRISTOFF et al. [C 37].

Substrates or substrates with inhibitors: (A) acetylthiocholine iodide Ac, (B) butyrylthiocholine iodide, Bu, (C) Ac + 62 C 47, (D) Ac + iso-OMPA, (E) α-naphthylacetate, Na. 1 = cholinesterase, 2 and 3 = acetylcholinesterase. Per gel: 0.1–0.2 ml extract; gel system No. 1, 2 cm spacer gel. Electrophoresis: 3 mA/gel at 4° C. For thiocholine ester, substrate incubation according to Karnovsky [K 6]: Concentration 5 mM, 25° C, 6–12 hrs. One hour of preincubation with inhibitor solution for the inhibitors.

In metachromatic leukodystrophy, a pronounced regression of *cerebroside sulfatase* activity is observed. MEHL and JATZKEWITZ [M 41] demonstrated by disc electrophoresis and other methods that the deficiency of a thermally unstable enzyme component with arylsulfatase type A activity is responsible for this condition. KASCHNITZ [K 7] reported on isozymes of *prostatic acid phosphatase* and SUSSMAN et al. [S 84] on the existence of organ specific *alkaline phosphatase isozymes* in liver and placenta. Changes of *lactate dehydrogenase isozyme* patterns were found by LANGVAD et al. [L 6] in colon tumors, KNUDSON and GORMSEN [K 23] in thrombocytes from acute phlebothrombosis, LEISE et al. [L 16] in leukocytes as an indicator of infection prior to overt symptomes and by RABINOWITZ et al. [R 2–3] in lymphocytes from chronic lymphatic leukemia. See also disc tissue electrophoresis page 68.

6.4. Other Tissue Proteins

RICHTER [R 37–38] found characteristic patterns of different *ferritins* in human, horse, rat, HeLa and KB cells. NEWCOMBE and COHEN [N 36] as well as GLENNER and BLADEN [G 11] investigated structural proteins of *amyloid fibrils* in human amyloidosis liver, FRIESEN et al. [F 42] the biosynthesis of placental proteins and lactogen and LISSITZSKY et al. [L 33] *thyroglobulin* from congenital goiter with hypothyroidism.

6.5. Epidermal, Hair and Nail Proteins

Even hard, keratinous tissues, such as hair and nails, contain at least two proteins which are soluble in Tris buffer (pH 8.5), as demonstrated by MATOLTSY [M 23] by disc electrophoresis of the pulverized horny material (3–5 mg) (for the method, see p. 59). Calluses of the feet and of the legs produced 4 and 5 bands, while pathological material in exfoliative dermatitis and psoriasis contained 7 and 8 soluble proteins, respectively. In addition, MATOLSKY [M 22] found that up to 10 proteins can be extracted from the three different layers of the cow nose epidermis (stratum corneum and granulosum, str. spinosum and str. basale); the solubilities of these proteins decrease from the top downward; this observation is of interest in view of the well-known differences in the degree of differentiation of individual cutaneous cell layers. Keratinous epidermal proteins can also be solubilized by trypsin or chymotrypsin for disc electrophoretic investigation [R 56]. Finally, studies of clinical interest were reported on hair kerateines [S 37] and noncollagenous proteins of human dermis [F 27].

7. BIOCHEMICAL APPLICATIONS

7.1. Enzymology

For the method of disc enzyme electrophoresis, see page 102.

7.1.1. Oxidoreductases

As shown by GOLDBERG, excellent separations can be obtained of *lactate and malate dehydrogenase* isozymes (LDH and MDH) from human and animal spermatozoa [G 15–16], testes [G 15], blood [G 15], snails [G 19] and trout tissue [G 17–18], with the use of gel system No. 1 (Fig. 71). The same is true for the isozymes of MDH from fungus-infected bean leaves [S 64] and sea urchin embryos [B 26, M 63], which were studies on morphogenetic problems. RABINOWITZ and DIETZ [R 1] performed similar analysis for the genetic control of LDH and MDH isozymes in lymphocyte cultures, while HATHAWAY and CRIDDLE [H 13] isolated beef heard LDH isozymes preparatively. Another example for a successful preparative purification was performed with thioredoxin reductase from E. coli extracts [T 9].

RUSSELL et al. [R 62] found that commercial preparations of *L-aminoacia oxidase* from snake venom may differ considerably. MITCHELL and WEBER [M 52] found three activators and one proenzyme for the *phenoloxidase system* in *Drosophila* pupae and observed by disc electrophoresis (gel system No. 1, 6.5 % separation gel) how the combination of one activator furnished tyrosinase activity with the proenzyme, while the product of the other two activators resulted in dopa-oxidase activity.

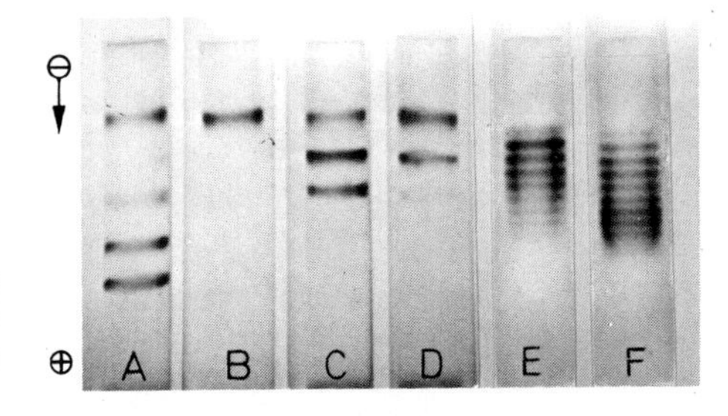

Fig. 71. – Lactate dehydrogenase isozymes from mouse testes (A–D) and trout tissue (E and F). Acc. to GOLDBERG [G 16–17].

(A) Lactate as substrate, (B) DL-α-hydroxyvalerate as substrate (C), isoenzyme recombinant (by a freeze-thawing process) with lactate as substrate, (D) recombinant with hydroxyvalerate as substrate, (E) LDH-isozyme from splake muscle, (F) from the heart of the hybrid of Salvelinus namaycush ♀ × S. fontinalis ♂.

The method has been suited for the identification, characterization and purity verification of *lactate dehydrogenase* from rabbit and bovine ocular lenses [F 31] and from *Bacterium subtilis* [Y 3], of 3α- and *3β-hydroxysteroid oxidoreductase* from rat liver [D 27], an *NAD-dependent aldehyde dehydrogenase* from bovine liver [R 47], *glyceraldehyde-3-phosphate dehydrogenase* from *Bacillus stearothermophilus* [A 27] and from mammalian mitochondria [F 18], a renal *α-hydroxy acid oxidase* [A 15], and of *tryptophanase* from *Bacillus alvei* [H 30], etc. SKYRING et al. [S 44a] developed an improved detection method for *bacterial dehydrogenases.*

CLARE et al. [C 39] provided evidence for the induction of *peroxidase* in sweet potato roots by disc electrophoresis, while FELBERG and SCHULTZ [F 11] reported on *myeloperoxidase*. TAKAYAMA et al. [T 3] used a highly acid continuous gel system for the fractionation of four electron-transporting enzyme complexes from mitochondria. FLATMARK and SLETTEN [F 26] studied polymorphism and turnover of *cytochrome c* from various rat tissues. For the detection of *enzymes converting nonreducing substrates to reducing products* (e.g. sugars) see page 103.

7.1.2. Transferases

FREDRICK reported on the identification and properties of isozymes of *glycogen phosphorylase* [F 38], *amylomaltase* and *α-glucan-6-glucosyltransferase* [F 37, F 39–41] from algae, while YUNIS and ARIMURA [Y 7] investigated the isozymes of glycogen phosphorylase from rat tissue.

Preparations of the following enzymes were tested for their purity: e.g. *phosphorylase a and b* [H 43], *carbamate kinase* [B 33], *DNA-polymerase* [F 2, N 28], *RNA-polymerase* [K 34, N 31–34], *ribonucleases* [L 15, P 22, W 27]. For rapid and sensitive detection methods of RNAse after polyacrylamide gel electrophoresis see e.g. [W 20, W 27].

7.1.3. Hydrolases

ALLEN et al. [A 23] found up to 19 *esterases* in mouse plasma and determined them by quantitative densitometry. Thus, sex associated differences could be quantitated [A 21]. The technique represents an advance compared to starch gel electrophoresis: Starch gel becomes transparent only by treatment with glycerol at elevated temperature [M 18]; in addition, the quantity of dye formed in the polyacrylamide gel is directly proportional to the incubation period (up to 15 min.). CENTER et al. [C 24] demonstrated the presence of genetic control of the luxoid gene over mouse liver esterases using quantitative densitometry, while DESBOROUGH and PELOGUIN [D 15] found similar genetically controlled differences in the isozyme patterns of potato esterases. FISH et al. [F 22] showed that disc electrophoresis can be used to identify and characterize tissue cells and their clones in suspension culture by their specific esterase isozyme patterns (Fig. 72). These patterns were found to be stable through consecutive subcultures.

Comparing four different methods of detection for *alkaline phosphatase* in starch and polyacrylamide gels, ALLEN and HYNCICK [A 17] found that the lead conversion method as well as the azo-coupling method has a higher sensitivity in polyacrylamide gel than in starch gel (sensitivity in standard polyacrylamide gel: 3 nM with the lead method and 1.5 nM phosphate with the azo-method).

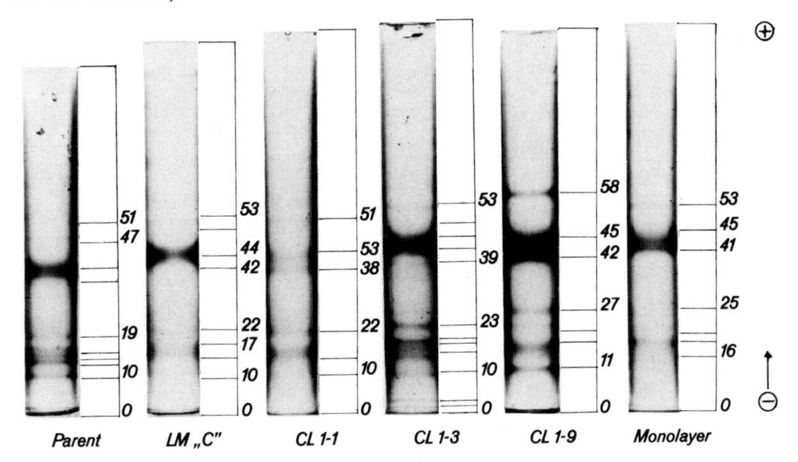

Fig. 72. Characteristic zymograms for a-naphthylbutyrate esterases from the LM cell line (parent), its clone Cl 1 and subclones Cl 1–1, 1–3 and 1–9. From: FISH et al. [F 22].

Gelsystem No. 1 with 8.5 % separation gels (TEMED concentration doubled), spacer gel, sample containing 15 % sucrose and cap gel (spacer gel).

CUNNINGHAM and FIELD [C 56] reported on the isozymes of alkaline phosphatase in serum, brain and liver of normal guinea pigs and those with allergic encephalomyelitis, while BARKA [B 6–7], ALLEN and GOCKERMANN [A 16] and ANDERSON et al. [A 29] described the separation and characterization of the isozymes of *acid phosphatase* from rat liver and brain. Acid phosphatase isozymes are also found in gonad homogenisates of silk worm pupae [G 8], in rabbit alveolar macrophages [A 47], as well as in bean leaves [S 63] which in addition contain isozymes of different esterases [R 59]. The method by AXLINE [A 47] utilizes a detergent (Triton X-100) for enzyme extraction and yields excellent results. RUSSELL et al. [R 61] used the method to test the purity of different preparations of *snake venom phosphodiesterase*, LERCH [L 17] to study multiple forms of plant *phosphodiesterase I*, TONO and KORNBERG [T 19] for the preparative purification of *inorganic pyrophosphatases* and YU et al. [Y 6] for the preparative isolation of *keratinase*.

In the above studies, the enzyme activities were localized by incubation of the gels with substrate solutions at optimal pH and temperature conditions. However, for enzymes with high molecular weight substrates, this method is not suitable. In such cases, it is advisable to carry out the localization with substrate containing separation gels (see page 104). BOYD and MITCHELL

[B 50] used this method for the identification of *deoxyribonucleases*, DOANE [D 23–24] for *amylases*, STEGEMANN et al. for *amylases*, *phosphorylases* [S 41 a], *polygalacturonases* (*pectinases*) [S 66 a] and *hyaluronidases* [L 30a]. BOYD and MITCHELL [B 50] obtained a quantitative reproducibility of ± 5 % with a limit of detection of 0.25 nanogram enzyme. DOANE later preferred the contact print method (page 106) [D 25–26] for the quantitative determination of *amylase* isozymes.

ZELDIN and WARD [Z 3] studied the isozyme patterns of an *α-amylase* in the course of the development of myxomycetes. MUUS and VNENCHAK [M 73] reported on the isozymes of salivary amylase. Both groups of investigators determined the enzyme activities in the eluate of individual gel fractions; ZELDIN and WARD [Z 3] obtained a nearly quantitative recovery of the enzyme activities. DEPINTO and CAMPBELL [D 13] reported on the purification and preparative isolation of amylase.

Disc as well as disc immunoelectrophoresis are suited for the identification, characterization and purity test of *sialidase* (*neuraminidase*) according to SETO and HOKAMA [S 32]. Studies on the chemical structure of subunits of *β-galactosidase* were carried out by several authors [C 53, G 10, K 32]. The *β*-galactosidase of uninduced, lactose metabolizing E. coli strains exhibits uniform electrophoretic properties [A 35] but appears in at least 7 enzyme forms after enzyme induction.

Disc electrophoresis was used as purity test for numerous enzymes: e.g. *invertase* [M 44], *carboxypeptidase-A* forms [F 29–30]; a zinc-containing particle-bound kidney *dipeptidase* [C 6]; a *peptide hydrolase* from fibroblasts [S 28]; *lysozyme*, *papain*, *chymopapain* [C 22]; *trypsin* and *chymotrypsin* [R 23]; *acetylcholinesterase* [L 20] etc. GUILBAULT et al. [G 46a] developed a rapid and sensitive assay of *cholinesterase* in polyacrylamide gels.

7.1.4. Lyases, Isomerases and Ligases

According to TAPPAN et al. [J 1, F 4] the isozyme patterns of *carbonic anhydrase* from the erythrocytes of newborn and adult men, cattle and guinea pigs differ considerably. HENNING and YANOFSKY [H 19] reported on mobility differences of the *tryptophan synthetase A*-proteins in 9 single and 3 double mutants of E. coli by amino acid exchange. For preparative studies see [D 33]. ROSENTHALER et al. [R 55] used the method to check the purity of a *histidine decarboxylase* preparation, while STRUIJK and BEERTHUIS [S 77] employed it for the separation of a protein fraction with *vinylacetyl-CoA-Δ^3-Δ^2-isomerase* activity. BERG et al. also applied it to test the purity of *tyrosyl-* [C 5] *and isoleucyl-* [B 3] *tRNA-synthetase* and ARONSON and DAVIDSON [A 38] to *hyaluronidase*. CHAN et al. [C 29] found by disc electrophoresis that *muscle aldolase* consists of non-identical subunits. Moreover *l-histidine ammonia-lyase* [R 17] and *maleate isomerase* [S 11] were purified on a preparative scale. Interestingly, on analytical disc electrophoresis, the purified isomerase showed, besides the active enzyme, an inactive, denatured protein produced during the analytical purity-checking run.

HASSALL et al. [H 12a] developed sensitive in situ detection methods for *histidase* and *urocanase* on polyacrylamide gels.

7.2. Endocrinology

7.2.1. Pituitary Hormones

Lewis et al. used disc electrophoresis to observe the influence of proteinases of the anterior pituitary lobe [L 22], of natural and synthetic proteinase inhibitors [L 23, L 29–30] as well as of different pH values and urea molarities [L 25] on the fragmentation of bovine *growth hormone* (*somatotropin*, *STH*) and *prolactin* (*luteotropin hormone*, *LTH*); they found a relation between the serum concentration of pituitary proteinase inhibitors and the status of the thyroid [L 30]. In addition, these authors observed that estradiol in ovariectomized rats stimulates prolactin production, while estradiol and cortisol in hypothy-

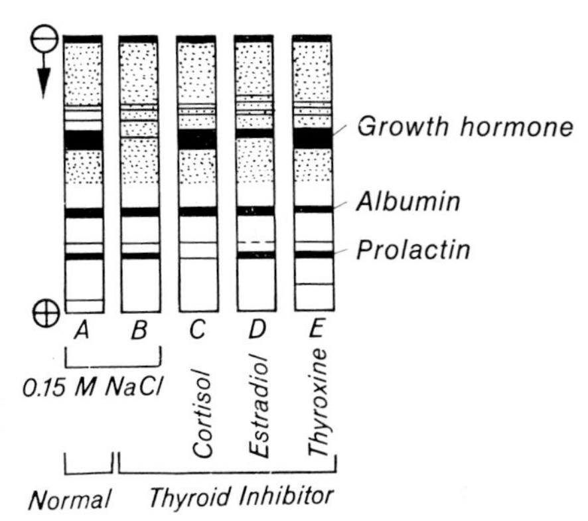

Fig. 73. – Influence of cortisol, estradiol and thyroxine on the protein pattern of the pituitaries of male, hypothyroid rats. From: LEWIS et al. [J 15].

Propylthiouracil as thyroid inhibitor and the steroid hormones were applied simultaneously for 3 weeks. Thyroxine for 7 days. In hypothyroidism, the growth hormone concentration decreases extensively (B); cortisol (C), estradiol (D) and thyroxine (E), however, restore the normal condition (A); the prolactin concentration increases more with estradiol administration (D) than with cortisol (C). – Gel system No. 1.

roid rats stimulate the production of growth hormone [J 15] (Fig. 73). Finally, they found that growth hormone is absent in the pituitaries of dwarf mice [L 27]. MEISINGER et al. [M 42] and REISFELD et al. [R 24] used the method to test the purity of somatotropin preparations, while SAXENA et al. [S 6] and LEWIS et al. [L 26] employed it for preparative purposes (see also [P 12] and [S 61]). LEWIS et al. [L 26a] noted a remarkable influence of fatty acids on the electrophoretic behavior of several pituitary hormones (prolactin, growth hormone, thyroglobulin; see also page 67). In the process of purifying bovine growth hormone Reusser [R 32] found that disc and disc immunoelectrophoresis offer better purity criteria than starch gel electrophoresis, N-terminal acid analysis and biological activity. Similarly Reisfeld et al. [R 26] compared paper, starch gel and disc electrophoresis during the purification of sheep prolactin. Yanai and Nagasawa [Y 2] found the method reliable for quantitative determination of prolactin. Purified fractions of *follicle-stimulating hormone* (*FSH*) [H 12], *α-melanocyte-stimulating hormone* (*MSH*) [R 23] and of *parathyrotropin hormone* (*PTH*) [H 14] were tested for homogeneity by disc electrophoresis. Human FSH and LH were also purified on a preparative scale [S 7]. In addition, reports have been published on a natural, biologically active peptide from sheep pituitaries [B 27] which resembles adrenocorticotropic hormone (ACTH), and on the proteins of the anterior and posterior lobe of the pituitary [R 27, R 48]. Smeds [S 48a] used microdisc electrophoresis to analyze soluble thyroid proteins and the colloid extracted by micropuncture from one rat thyroid follicle.

7.2.2. Insulin

Heideman [H 16–17] used the ^{131}I-labelled proteohormones *insulin* and *thyrotropin* to study the patterns and properties of soluble antigen-antibody complexes, while Yagi et al. [Y 1] investigated the differences in the antigenicity behavior of ^{125}I-insulin and its two chains. Bauer et al. [B 11] reported on the biosynthesis of insulin. Chance et al. [C 30] demonstrated how trypsin cleaves insulin and proinsulin to different polypeptide fragments during the course of digestion. Kvistberg et al. [K 42] described a test for insulin identification with aldehyde fuchsin.

7.3. Immunology

Searching for a possible relation between antibody-specificity and electrophoretic behavior, Reisfeld, Small et al. [R 22, R 25, S 47–48] investigated *H- and L-polypeptide chains* of rabbit *γG-immunoglobulins* and *haptene-specific antibodies* (gel system No. 11). The polypeptide chains could be separated into certain electrophoretic subgroups in the presence of urea; however, different specific antibodies and normal γG-immunoglobulin could not be distinguished, a fact which is attributable to the pronounced heterogeneity of the antibody molecules [R 25]. Similar findings were obtained by Roholt et al. [R 51–52] (gel system No. 1a and 7). Moreover, Reisfeld et al. [R 22a] isolated γG-L-chains on a preparative scale. Stelos et al. [S 69] reported on multiple molecular forms of purified antibodies from rabbit serum; a sharp separation of the components could be obtained only with disc electrophoresis (gel system No. 1).

7.4. Animal Body Fluids

7.4.1. Animal Milk Proteins

AKROYD [A 12] gave a survey on the disc patterns of milk proteins (casein and whey protein components). Disc electrophoresis provided a powerful method to separate *lactotransferrin* (*red protein*), *lactoperoxidase*, *α-lactalbumin* [G 44], *κ-casein* [W31–32] and to study polymorphism of *γ-casein* in cow's milk [G 45]. Moreover, its usefulness in food technology to monitor protein changes reflecting e. g. aging processes is now well documented (see page 185).

7.4.2. Animal Serum Proteins

NARAYAN et al. [N 15] separated animal serum proteins from chicks, rats, mice, rabbits and guinea pigs. Bovine *transferrin* types can be determined better by disc than by starch gel electrophoresis [R 9]. The method also proved useful to check the purity of the isolated *bovine Hageman factor* [S 17–18]. More than 30 fractions were found in bovine and sheep serum by discontinuous gradient gel electrophoresis (p. 62) [W 36]. In addition, the serum proteins of pigs, goats, cats, grasshoppers [A 32], of different pigeon species [D 14], blood and coelomic proteins of Lumbricus terrestris and Glycera dibranchiata [L 32] and *growth factors* of trypanosomes from rat serum [G 33] have been investigated. CHADER and WESTPHAL [C 25] demonstrated the polymeric nature of a *corticosteroid-binding globulin* of rat serum, thus providing evidence for the regulatory control by a steroid hormone over the quaternary structure of a steroid-binding protein. *Lipoproteins* (from rat serum) can also be advantageously separated by disc electrophoresis [D 36, N 10–11, N 13] and the same is true for *mouse myeloma proteins* [G 21].

CHEN [C 32] reported on the serum proteins of different species of amphibia, COATES and TWITTY [C 44] on salamander proteins. MOSS and INGRAM [M 70] demonstrated that *tadpole hemoglobin* (T-Hb) biosynthesis by erythrocytes rapidly diminishes under the influence of thyroxine; after hormone treatment for 15 days, however, the tadpoles have erythrocytes which produce frog hemoglobin (F-Hb) (Fig. 74). T-Hb and globinpolypeptide chains have then been separated on a preparative scale [M 71]. BARKER [B 9] similarly monitored the biosynthesis of embryonnic and adult hemoglobins of the mouse hematopoietic system.

According to PAYNE [P 7], ocean perch, pollock, haddock, flounder and cod fish have species-specific serum protein patterns which, if known, permit an identification of commercial fish filets (gel system No. 1). This points to a suitable application of the method in food technology and forensic analysis (see page 185). MANCUSO [M 5] extended the number of investigated fish species to twenty-two. DAVIS and LINDSAY [D 10] in addition were able to distinguish between subspecies of Oncomelania formosana. The analyses indicate that very specific protein patterns can be assigned to certain genera and species, while in subspecies, the patterns are mostly homologous but occasionally deviate in some proteins. Thus, *hybrids* can be readily identified (Fig. 75). Moreover, even age-dependent changes have been detected., e.g. in the blood protein pattern of the lobster [C 26].

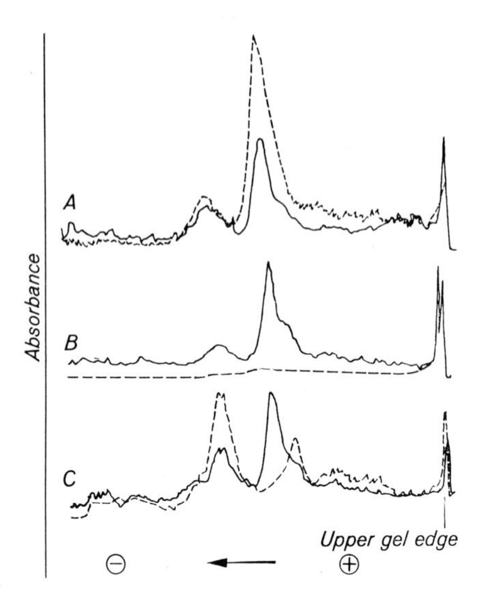

Fig. 74. – Densitometer traces of tadpole hemoglobin proteins. From: Moss and Ingram [M 70].

—— Amido Black stain, – – – autoradiogram after 2 hrs. of *in vitro* incorporation of C^{14}-algal hydrolysate in tadpole erythrocytes at 29° C. Electrophoresis of the reduced and alkylated heme-free globins in 8 M urea of (A) control animals, (B) animals 8 days after thyroxine treatment, (C) animals 15 days after thyroxine treatment. – Gel system No. 9, electrophoresis 1 hr. at 20° C and 0.5 mA/gel, then 2–6 hrs. at 3 mA/gel. Autoradiography according to Fairbanks et al. [F 1], p. 100. Densitometry of the autoradiograms.

7.4.3. Other Animal Body Fluids

Kunitake et al. [K 38] investigated the *uterine fluid* of rats by disc immunoelectrophoresis. Kirchner [K 20a] found uterus specific glycoproteins during early pregnancy in the uterine and blastocoelic fluid of the rabbit.

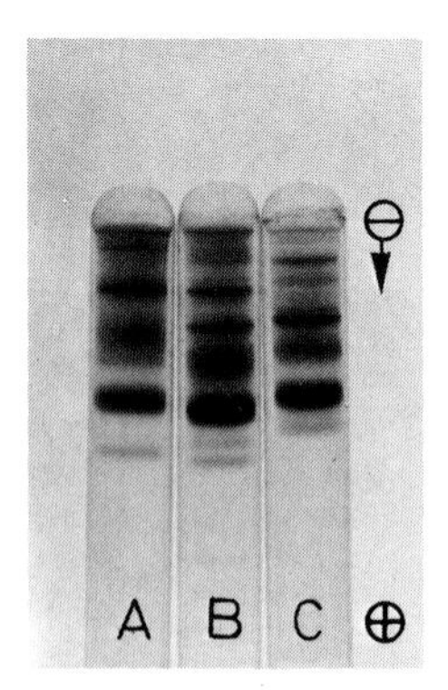

Fig. 75. – Serum protein pattern of Pennsylvanian sweet-water fish (J. E. Wright, Jr.). (A) Esox lucius ♀, (B) F_1-hybrid of Esox masquinongy ♂ × Esox lucius ♀, (C) Esox masquinongy ♂. Amido Black staining.

7.5. Neurology

PUN and LOMBROZO [P 32] developed a micromethod (page 111) to separate the proteins from small quantities (1–2 mg) of individual cerebral organs of rats. While 7–10 proteins could be identified by starch gel, agar gel and celluloseacetate electrophoresis, polyacrylamide gel electrophoresis permitted the detection of up to 18 fractions. HYDÉN et al. [H 49, M 37] used another micromethod for the analysis of specific brain proteins. An improved microtechnique has been developed for this purpose by NEUHOFF [N 27] (see page 112).

VOS and VAN DER HELM [V 8] found 22–25 protein fractions in the grey matter of rabbit brains using Triton X–100. NEET and FRIESS [N 22] tested chromatographic fractions of *curare-binding macromolecules* of medullated nerve tissues from frogs and dogs for homogeneity. The method was also used by NAKAO et al. [N 8–9] and KIBLER et al. [K 15] to characterize and check the purity of an *encephalitogenic* protein from bovine spinal cord. The axoplasm of the squid (Dosidicus gigas) was found to contain 7 proteins which are not present in the serum according to disc and disc immunoelectrophoresis [H 45] (gel system No. 1, a dialysis membrane prevented the sample material from diffusing into the upper electrode buffer). SMITH and VARON [S 49] demonstrated that the *nerve growth factor* (from mouse salivary gland) exists in multiple forms. Even brain tumors show characteristic protein patterns in disc electrophoresis [S 44b].

7.6. Histology and Cytology (Animal Tissue and Cell Proteins)

7.6.1. Crystallin and other Ocular Lens Proteins

Crystallin and its subunits can be separated and characterized better by polyacrylamide gel than by starch gel electrophoresis according to BLOEMENDAL et al. [B 37–38] and WISSE et al. [W 24]; these authors emphasized the controllability of the pore size as a special advantage for their purposes. BJÖRK [B 34] observed different electrophoretic mobilities, but similar immunological properties of α-crystallin isolated from 8 mammalian species. KONYUKHOV and WACHTEL [K 31] studied changes in the crystallin pattern of normal lenses and cataracts of inbred and mutant mice. YOUNG and FULHORST [Y 4] were able to observe changes in protein biosynthesis with cell specialization in the lens cells of rats. See also [R 18].

7.6.2. Collagen, Muscle and Cutaneous Proteins

The studies of a number of authors [B 19, B 46–47, F 36, G 39, K 4, N 3, P 20, V 3 etc.] on the protein chemistry of *collagen* and its peptide chains indicate that disc electrophoresis (usually gel system No. 8) has become indispensable as a rapid and reliable method for the following purposes:

evaluation of extraction methods, purity tests of chromatographic fractions, investigation of the conversion of α- and β-chains, observation of cleavage reactions and inter- and intramolecular polymerizations of subunits, determination of quantitative ratios in the chains after different fractionation methods and for the study of enzyme reactions and pathological processes expressed by changes in molecular structure.

In order to determine the composition and structure of different collagen chain aggregates, VEIS and ANESEY [V 3] developed a special staining method. BENSUSAN [B 19] used the same method to study products which form during the iodization of tropocollagen. For the identification of collagen components with a high mobility, FESSLER and BAILEY [F 16] subjected the gels to preelectrophoresis (p. 59). The resolving power of polyacrylamide gel electrophoresis in this case exceeded that of analytical ultracentrifugation. Using a three layer discontinuous gradient gel (7.5, 5.5 and 4.0 % gel layers) CLARK and VEIS [C 40] were able to resolve the α-, β-, and γ-components as well as a pre- and high-polymer region of bovine corium collagen. For exact comparison the split gel technique was used (see page 64).

GRAHAM and GILLIBRAND [G 31] studied *actin polymers*, CAHN [C 1] investigated the immunochemical properties of *tropomyosin* and other actin-associated antigens. See also [W 28]. ARAI and WATANABE [A 36] concluded from their disc electropherograms that the *relaxing protein* is a complex of *troponin* and *tropomyosin*. For *actomyosin* see [A 45]. FLORINI and BRIVIO [F 28] used dilute gels (containing 2.6 % acrylamide and 9 M urea) to fractionate the large subunits of *myosin*. According to ADELMANN et al. [A 7] non-collagenous proteins from the stroma of cutaneous connective tissue can also be readily determined by disc electrophoresis.

7.6.3. Histones and other Nuclear Proteins

Histones are assumed to be functionally involved in a specific repression of genetic activities. To test this idea, particularly sensitive methods are required. Since disc electrophoresis offers a high resolving capacity and needs only small quantities of histones, this method has been increasingly preferred by a number of authors to study problems of morphogenetics and cell differentiation. Thus, no tissue specific, but species-specific and age-dependent histone differences were observed in several rodent tissues [N 23]. However, this was not confirmed for chick embryo histones [K 21]. Similarly histones from mammary glands at different stages of development and lactation were analyzed [S 68]. Moreover, histones from dividing and non-dividing cells have been compared [H 8, L 14].

These and other studies of histones biochemistry have been made possible by the introduction of appropriate gel systems. The first suitable gel system (No. 7, p. 44/45) for basic proteins using discontinuous buffers, was developed by REISFELD et al. [R 23] and prompted, partly after slight modification, numerous studies, for example on

pea bud nucleohistones [H 42], degradation products of a lysine-rich histone from calf thymus [K 20, R 20], mammalian brain, liver and kidney histones [N 23, P 21], histones from mammalian chromosomes [M 1b], on the biosynthesis and metabolism of histones in HeLa cell nuclei [S 61a] and cytoplasm [R 46], in calf endometrium, rat liver, pea cotyledons and tobacco cells [C 28], mouse brain and liver [P 21].

Histones tend to aggregate in sample and spacer gels, particularly when the pH discontinuity is pronounced. MCALLISTER et al. [M 34] therefore dispensed with sample and spacer gels and applied the histone solution directly on the separation gel, in a 1 : 4 (v/v) dilution of the electrode buffer to which sucrose has been added to a final concentration of 1 M. This provided thin starting zones due to the conductivity jump (see page 25). In contrast to this continuous system, SHEPHERD and GURLEY [S 40, G 48] explored the potentials of discontinuous buffer systems to produce a steep voltage gradient (gel system No. 10, pp. 46/47) and achieved improved resolution.

Systems omitting spacer and sample gel were adopted by several workers who partly included *urea* for better resolution [for example B 12–13, B 44, C 28, F 3–6, H 42–43, L 14, M 19, M 32, O 6–7].

Thus FAMBROUGH and BONNER [F 6] found that arginine-rich histones, isolated preparatively from pea buds and calf thymus, show striking similarities, suggesting that histones of plants and animals remained unchanged over the course of evolution in higher organisms.

PANYIM and CHALKLEY [P 3] thoroughly examined several experimental conditions (current, preelectrophoresis [see page 59], gel and urea concentration, pH) to optimally resolve the five major histone classes from calf tissues.

The maximal number of bands are found in gels containing 15 % acrylamide, 0.1 % Bis, 2.5 M urea and 0.9 N acetic acid pH 2.7, after electrophoresis for 9 hrs. at 1.75 mA/0.6 cm diameter gel and 115 V. For high resolution, very long gels (up to 25 cm) are used [P 4]. Yet the authors stress that no single set of experimental conditions can suffice to maximally resolve all histone bands.

This technique enabled the authors [P 4–4a] to detect a tissue specificity of histones and differences in dividing and non-replicating tissues (Fig. 76).

Disc electrophoresis is also suited for the analysis of other nuclear proteins, such as the *nuclear residual proteins* [D 29], or *acidic proteins* isolated from mammalian nuclei [B 17].

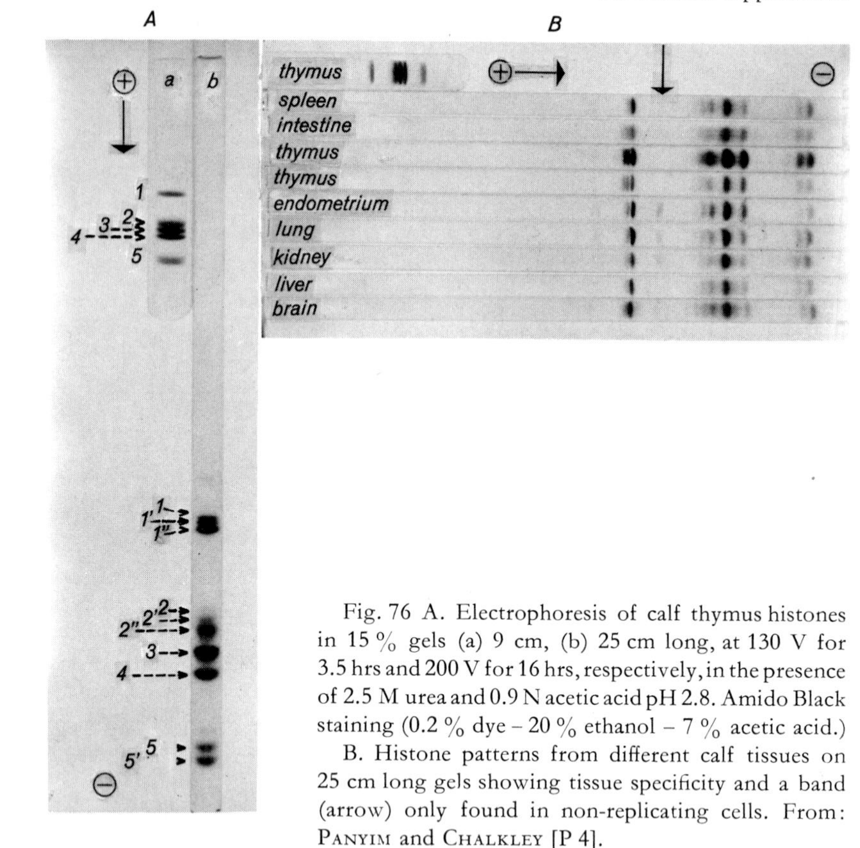

Fig. 76 A. Electrophoresis of calf thymus histones in 15 % gels (a) 9 cm, (b) 25 cm long, at 130 V for 3.5 hrs and 200 V for 16 hrs, respectively, in the presence of 2.5 M urea and 0.9 N acetic acid pH 2.8. Amido Black staining (0.2 % dye – 20 % ethanol – 7 % acetic acid.)

B. Histone patterns from different calf tissues on 25 cm long gels showing tissue specificity and a band (arrow) only found in non-replicating cells. From: Panyim and Chalkley [P 4].

7.6.4. Other Subcellular Proteins

Mammalian ribosomal proteins have been successfully separated both analytically [L 41, O 11] and preparatively [H 4].

In a comparison of the patterns of water-soluble proteins and esterases from normal skin and melanomas of the sword tail, the platy fish and their hybrids, Humm and Sylvia [H 44] found several tumor specific proteins.

7.6.5. Membrane Proteins

Several authors [C 42 K, 1, R 58, S 1–2, W 6] successfully fractionated *bacterial membranes*, partly after solubilization with dissociating agents (see page 67). Schneiderman [S 15–16] and Marchesis and Steers [M 9] separated *erythrocyte membrane proteins*, Neville [N 35] alkali-soluble proteins from *liver cell membranes*, using the modified gel system No. 9 with urea, while Manganiello and Philips [M 8] reported on the electrophoresis of *rough and smooth surface membranes* of isolated rat liver microsomes. BERKMAN et al. [B 21] found that urea greatly enchances the solubilization and electrophoretic resolution of the *microsomal* (*endoplasmic reticulum*) *membrane proteins* from rat liver.

Many proteins can only be liberated from membranes by the attack of strong dissociating agents. Thus, *mitochondrial membrane enzymes* [T 3] and brain proteins [C 51] were solubilized and electrophoresed by the "*PAWU*" *system* (phenol-acetic acid-water, 2 : 1 : 1, w/v/v made 5 M in urea). The similar "*PAMU*" *system* (phenol-acetic acid – 0.2 M mercaptoethanol, 2 : 1 : 1, w/v/v, made 5 M in urea) was used to dissociate *soy been proteins* [C 17]. Both systems do not need spacer gels.

7.7. Entomology

Bee, wasp, and hornet venoms exhibit similar but not identical protein patterns according to O'Connor et al. [O 1]; on the other hand, BENTON and PATTON [B 20] reported that geographic and seasonal differences cannot be detected. Such findings are of importance for the immunization of hypersensitive individuals. Other disc electrophoretic research has been conducted on the hemolymph *lipoproteins* of silk-moth pupae [C 33], *hemolymph* proteins of Cecropia pupae in the diapause [B 22], muscle proteins of different Hymenoptera and an Orthoptera species [J 13], different *enzyme patterns* in Drosophila melanogaster [B 50, D 26] (Fig. 43 and 44), turnover of Drosophila *hemolymph proteins* [B 52] and on proteins in the hemolymph, salivary gland and salivary secretion of Chironomus larvae [D 32, G 42]. The method was also suited for the characterization of *bursicon*, a proteohormone controlling cuticle tanning [F 35].

7.8. Botany

7.8.1. Seed Plants

Since genetic differences are often reflected in shifts of protein patterns, more refined separation methods, such as disc electrophoresis, are of special interest for research on the *taxonomy*, *biochemistry*, *physiology*, *phylogeny*, *morphogenesis*, *genetics* and *cultivation* of plants. This was pointed out by HAGMAN [H 1] in a study of protein patterns and immune reactions of electrophoretically separated substances from the pollen and catkins of Betula verrucosa, B. pubescens, and B. Medwediewi (gel system No. 1). He found species-specific and, by disc immunoelectrophoresis, even individual-specific differences in the pollen of these birch species as well as age-dependent differences in the catkins. Problems of autosterility can thus be investigated in greater detail.

The application of disc electrophoresis as a method for solving *taxonomic* and *phylogenetic problems* was also emphasized by Fox et al. [F 32–33] and VAUGHAN et al. [V 2] who investigated the seed proteins of legumes and Brassica species. The morphological classification thus can be supplemented by biochemical categories. Fig. 77 shows that the genera Rhynchosia and Phaseolus, which belong to the subfamily of Papilionideae, yield similar patterns but clearly differ from Caesalpinia gilliesii which belongs to the subfamily

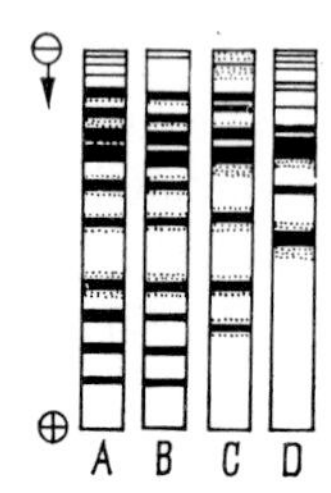

Fig. 77.– Seed proteins of different legumes. From: Fox et al. [F 33].

(A) Phaseolus vulgaris; (B) Phaseolus coccineus; (C) Rhynchosia phaseoloides; (D) Caesalpinia gilliesii. Gel system No. 1, 200 µg protein in each case.

of Caesalpinoideae. It is even possible to identify different genotypes of one species by their root protein patterns, as demonstrated by COLEMAN et al. [C 49] in alfalfa. The six investigated genotypes differed in the degree of their cold resistance. Similarly, the protein patterns of the embryo and female gametophyte of the Jack Pine differ according to climate at seed source [D 42]. Moreover JOHNSON et al. [J 9–11] demonstrated that, by quantitatively evaluating protein patterns of *wheat* species, useful taxonomic standards can be set up to identify the different types. *Fruits* also exhibit characteristic protein patterns, as observed by CLEMENTS [C 43] in apples, pears, oranges, lemons, bananas, and in avocados. Since the protein content of these fruits is small and homogenizing of these fruits liberates plant acids and polyphenolic compounds, which partially denature the proteins, the author selected a mild low-temperature extraction method.

The biochemical mechanism of *dormancy* was investigated by PERRY [P 13] in Spirodela polyrrhiza. The considerable differences of the protein patterns of the vegetative fronds, the turion-forming fronds as well as of the turions themselves indicated that protein synthesis is inhibited in dormant tissues.

STEWARD et al. [B 4–5, C 31, S 71–73] also employed disc electrophoresis for research on *morphogenetics* of Neurospora crassa, Pisum sativum and Tulipa. The method furnished superior separations compared to paper and starch gel electrophoresis. It proved to have a satisfactory reproducibility not only for identical but also for different protein extracts of identical origin and permitted the evaluation of different extraction methods with respect to their yields and denaturation effects [S 71]. Photopolymerized gels produced a sharper separation than persulfate polymerized gels [F 69].

In addition, the distribution of Tris-soluble proteins in different segments of the growing pea seedling was followed and was found to be an expression of varying gene activities [S 73].

Various organs of the tulip at different growth stages exhibit organ-specific protein, MDH and esterase patterns, which change with time; the protein changes precede the morphogenetic responses [B 4]. By disc immunoelectrophoresis changes in the reserve proteins in germinating soy bean seeds could be detected [C 17a]. See also [C 17, C 17b]. Similar age-dependent differences were found in ivy [F 43] and in conifers [D 41–42].

However BESEMER and CLAUSS [B 23] expressed *warnings* as to the validity and significance of these studies. The protein patterns may be altered by external factors such as extraction method, electrophoretic conditions, gel concentration etc. Interactions of different proteins should be minimized. Denaturations due to the action of polyphenoloxidase may occur during tissue homogenization unless mercaptoethanol, polyvinylpyrrolidon or polyethylenglycol are present. Also the ratio of extraction medium to protein amount of the tissue plays a role. Furthermore associations of enzymes with different proteins may change their activities. Thus, even if the reproducibility of patterns is ascertained it remains to be proven that artificial reactions do not occur.

Polyacrylamide gel electrophoresis proved to be a sensitive tool for examining a great variety of plant proteins of interest for food science and technology, for example proteins from *sorghum flour* [S 5], *wheat flour* [N 18], *wheat leaves* [W 33], *wheat seeds* [J 9–11], *soybean leaves* [H 21], *soy bean seedlings* [C 17, C 17a–c], *soy bean whey* [E 3a], *potato tubers* [L 37a], *groundnut cotyledons* (*arachins*) [T 18a]. Molecular biologists used the method to analyze *phytohemagglutinin* [T 2], *wheatgerm agglutinin* [B 61], *proteins from spinach chloroplasts* [R 39], *basic ribosomal proteins* from different species [L 43], *allergen-proteins* from castor beans [M 68] etc.

7.8.2. Protozoa and Fungi

SCHIMKE et al. [S 13] found in crude extracts of Mycoplasma hominis arginine deaminase, ornithine transcarbamylase and carbamate kinase activities, which differed clearly from those of an extract of Mycoplasma laidlawii A. – In investigating the question of whether *morphological differences* of closely related protozoa species can also be detected biochemically, KATES and GOLDSTEIN [K 8] compared the protein patterns of Amoeba proteus strain BK, strain T Amoeba discoides and Chaos chaos by disc, disc enzyme and disc immunoelectrophoresis and observed significant differences only between Amoeba and Chaos. Disc electrophoresis also provides a highly resolving non-immunological method for analyzing proteins of both parasite (Plasmodium and Anaplasma) and host origin [F 19, S 59–60]. ALLEN [A 24] found variable enzyme activities as an expression of changing gene activities in

Tetrahymena pyriformis. TABER and SHERMAN [T 1] were able to distinguish oxidized yeast *cytochrome c* from the reduced component (Fig. 78). If the hemoprotein was denatured by brief heating to 100° C, an oxidized cytochrome c component was obtained which bound carbon monoxide (peak A) and clearly differed from the reduced form (peak B).

CODDINGTON et al. [C 45] investigated the mechanism of *genetic complementation* between two Neurospora mutants with "defect" glutamic dehydrogenases by disc, disc enzyme and disc electrophoresis combined with autoradiography. See also [S 83]. As demonstrated by DURBIN [D 40] with different species of Septoria, a disc-electrophoretic protein pattern furnishes information on the *variability* of these fungus species.

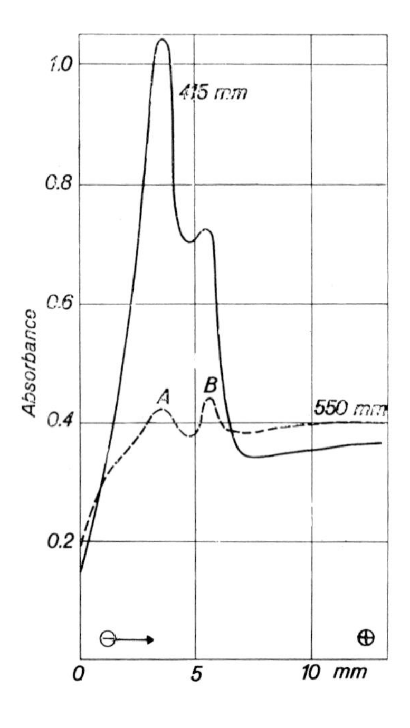

Fig. 78. – Disc electrophoretic separation of heat-denatured cytochrome c components from yeast. From: TABER and SHERMAN [T 1].

Spectrophotometric analysis with a micrometer device (Gel system No. 5).

7.9. Microbiology

7.9.1. Bacteria

In spite of immunological identity, proteins can have different mobilities in disc electrophoresis; this was demonstrated by FOX and WITTNER [F 34], in the case of the M-proteins of three serotypes of streptococcus group A (Fig. 79). – Remarkable examples of the application of disc electrophoresis in molecular biology are the detection, identification and purity test of a protein whose biosynthesis is controlled by one of the *three regulator genes* for *alkaline phosphatase* in Escherichia coli, i.e. the R2a gene. GAREN and OTSUJI [G 4] rapidly detected those phosphate-constitutive mutants which form the regulator gene product; their findings were confirmed and quantitatively refined by immunological cross reactions. In addition, by detecting the changed electrophoretic mobility of the revertant R2a protein, the authors were able to demonstrate that the R2a gene actually determines the protein structure (amino acid sequence) of its product. – Valuable contributions to our knowledge on the gene regulation of lactose-digesting enzymes of E. coli were made by the studies of CRAVEN et al. [E 53], STEERS et al. [S. 65–66] and of GIVOL et al. [G 10] in protein chemistry dealing with the properties, composition and structure of *β-galactosidase*. The authors were interested in the question of whether one of the functions of the lac-operator gene does or does not consist of directly controlling the formation of the proximal segment of β-galactosidase, or with other words, whether the operator gene represents a part of the structural gene or not. STEERS et al. [S 66] compared the properties of the isolated and purified β-galactosidase of a normal-constitutive strain of E. coli K 12 (3300) and an operator-constitutive mutant (0^c_{76}), and answered the question in favor of the latter possibility. These authors emphasized the value of disc electrophoresis to control

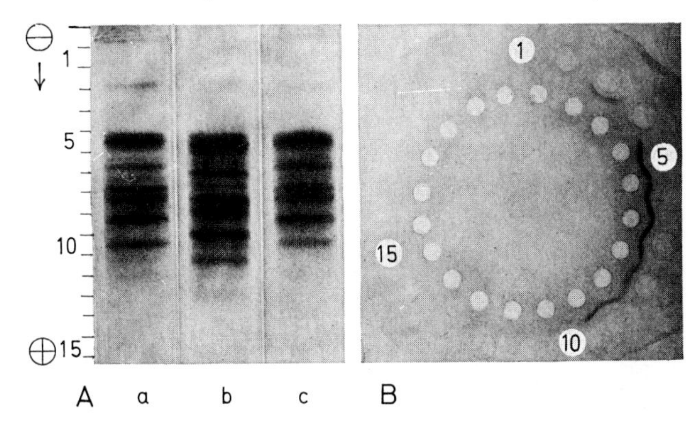

Fig. 79. – Disc electrophoresis and immunodiffusion of the M-antigen proteins (type 24) from streptococcus cell walls (group A). From: Fox and WITTNER [F 34].

The experiment demonstrated that the multiple bands do not represent artifacts due to covalently bound hexosamine and muramic acid oligosaccharides. In the presence of sucrose, different protein patterns would have formed.
A: (a) purified M-proteins, untreated; (b) purified M-proteins after incubation with lysozyme; (c) purified M-proteins after incubation with lysozyme and group-C-lytic enzyme. B: Agar immunodiffusion of the unstained gel fractions of A against absorbed homologous antiserum.

the efficiency of fractionation methods (Fig. 80) as well as its good reproducibility. TERZI and LEVINTHAL [T 8] used autoradiography to detect ^{14}C-labeled E. coli proteins following infection with λ-phages.

The *ribosomal proteins* of Escherichia coli can be separated into 28–34 protein bands in gel system No. 7 (with 8 M urea), representing about 85 % of all proteins (electrophoresis at 3° C and 3 mA/gel, 90–120 min). To permit a reliable comparison of such differentiated spectra, LEBOY et al. [L 12] introduced the *split gel technique* (p. 64) by which the K 12 strains of all other coli strains could be differentiated on the basis of a shift of the electrophoretic behavior of a single protein.

Fig. 81 shows parts from the complex protein spectra of a K 12 and a B strain. One protein (X) appears in strain K 12, while the concentration of another protein (Y) diminishes in strain B. The Y-protein (curve B) consists of 2 components: one is genetically varied in K 12 and appears as X-protein with a modified mobility. After separation of the ribosome subunits, it was possible to demonstrate that the Y-protein originates from the 30 S subunit.

Such electrophoretic differences are highly valuable for genetic research, because they permit an assignment of genetic characteristics to defined structural properties (*phenotype-genotype correlation*).

Disc electrophoretic separations of the *proteins* of the *ribosomal subunits* from E. coli have been described by several investigators [e.g. A 33, G 6, L 41, M 60, S 20, S 31, T 20–21]. Hjertén et al. [H 26] separated the intact 30 S and 50 S subunits by continuous polyacrylamide electrophoresis both analytically and preparatively. A continuous system was also used by Work [W 30] for the separation of the 30 S and 50 S ribosomal proteins from E. coli, 40 S and 60 S proteins from rabbit reticulocyte ribosomes and the proteins of crystalline encephalomyocarditis virus. – If ribosomes from E. coli are subjected to DEAE-cellulose chromatography, purified particles are obtained which contain about 30 % less protein than ribosomes treated by repeated centrifugation. However, according to Furano [F 45] the "DEAE ribosomes" retain their complete biological activity during chromatography (capacity to synthesize polypeptides and bind Phe-s-RNA in the presence of polyuridylic acid).

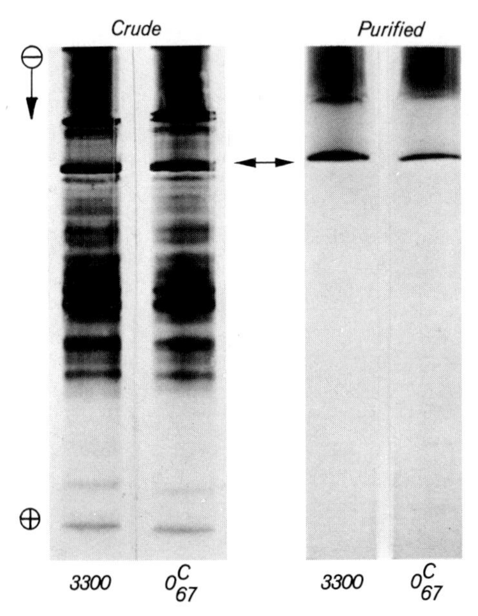

Fig. 80. – Disc electrophoretic protein patterns of β-galactosidase crude extracts and purified enzyme preparations from "3300", a regulator-constitutive strain of Escherichia coli K 12 ($i^-o^+z^+$) and "O^C_{67}", an operator-constitutive strain ($i^-o^cz^+$.) Arrow indicates enzyme activity. Gel system No. 1. From: Steers et al. [S 66].

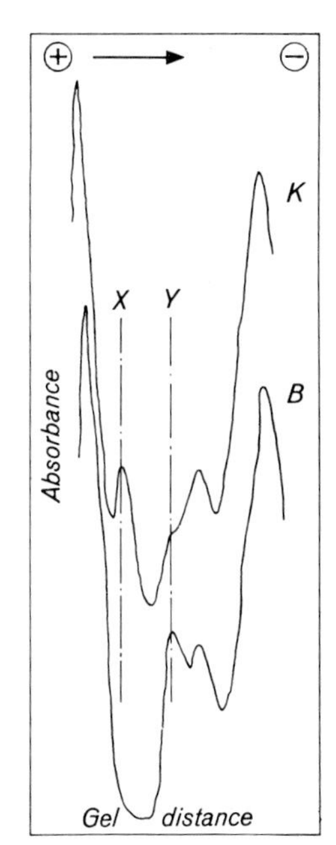

Fig. 81. Parts from microdensitometer scans of disc electrophoretic patterns of ribosomal E. coli proteins. From: LEBOY et al. [L 12].

Upper tracing K: proteins of strain K 10; lower tracing B: proteins of strain B. The dashed lines indicate the positions of the X- and Y-proteins. Gel system No. 7 with 8 M urea. Electrophoresis at 3° C and 3 mA/gel, 90–120 min. Pyronine Y as buffer front stain. Photography with Polaroid type 55 P/N film; evaluation of the negative with a microdensitometer of firm 17.

The protein loss becomes evident only in one protein band which cannot be responsible for it. This finding is remarkable because it demonstrates the caution that it indicated in the interpretation of such complex disc electropherograms. – Ribosomal proteins from baker's yeast were also fractionated on a preparative scale [H 40].

MARGOLIES and GOLDBERGER [M 15] isolated an isomerase from Salmonella typhimurium as the fourth *enzyme of histidine biosynthesis*. ASAKURA et al. [A 43] reported that the method is also suited for the analysis of thermally depolymerized *flagellins* of different Salmonella strains. FINKELSTEIN et al. [F 20] identified *choleragen* by disc immunoelectrophoresis. MONTIE et al. [M 62] observed the influence of different sulfhydryl reagents and denaturing substances on *toxins A* and *B* of Pasteurella pestis. SHEFF et al. [S 38] analyzed *tetanus toxin*, KALF and GRÉCE [K 3] *B-and C-antigens* from Erysipelothrix rhusiopathiae. Culture filtrates of Mycobacterium tuberculosis were found to contain not only proteins but also polysaccharide- and nucleic acid-containing components [A 8]. CHAIET and WOLF [C 27] succeeded in characterizing *streptavidine* as a new protein with a specific biotin-binding property obtained from the filtrates of Streptomycetes cultures. On *bacterial membrane proteins* see page 168.

7.9.2. Viruses and Phages

MAIZEL [M 2] separated the proteins of purified *type 1 poliovirus* into four components. Treatment with sodium dodecyl sulfate (SDS), urea and 2-mercaptoethanol (ME) yielded 14 components (4 coat- and 10 other virus-specific proteins) in a continuous gel system [S 82]. MAIZEL's [M 3] automatic gel fractionator (p. 93) proved to be especially appropriate here: It permitted the authors to follow the biosynthesis of virus-specific proteins (in HeLa cells) rapidly and with a high resolving capacity in numerous pulse-chase experiments.

The 10 % separation gels (10–20 cm × 0.6 cm) contain 0.1 % SDS, 0.5 M urea and 0.1 M phosphate buffer at pH 7.2. No spacer gel is used. Sample in 0.01 M phosphate buffer pH 7.2 containing 1–2 % SDS, 0.1 % ME and sucrose. Electrode buffer: gel buffer with 0.1 % SDS. Electrophoresis 4–8 hrs. at 3 V/cm.

Using a similar method, MAIZEL [M 3] found up to 10 different proteins in *adenovirus particles*. NATHANS et al. [N 19] observed 4 *phage-specific proteins* of E. coli lysates following infection with an RNA phage; for the quantitation of the C^{14}-labeled proteins, autoradiograms were prepared which could be evaluated with a microdensitometer. See also [S 79, V 4–5]. CELIS and CONWAY [C 23] demonstrated the in vitro biosynthesis of an early phage-induced protein using a T 2-DNA-dependent amino acid incorporating system. PTASHNE [P 31] succeeded in isolating a *λ-phage repressor* and in characterizing it by disc electrophoresis.

RUECKERT and DUESBERG [R 60] isolated mouse *encephalitis virus proteins* on a preparative scale. KONIGSBERG et al. [K 29] used gel systems Nos. 1 and 8 (with 8 M urea) to check the purity of a *f_2-bacteriophage coat protein* preparation. Mixtures of different capsid mutants of *bacteriophage* Φ χ 174 were readily separated by HUTCHISON et al. [H 48] in 2.6–5.3 % separation gels: the method made it possible to draw important conclusions concerning the mutants of capsid genes. KELLENBERGER [K 10] identified the capsid components of *phage T-even* following degradation. Van de WOUDE and BACHRACH [V 1] provided evidence for the existence of a single structural polypeptide in *foot-and mouth-disease virus*.

Intact viruses can also be isolated with a high degree of purity under mild conditions (i.e. without denaturation) by preparative polyacrylamide gel electrophoresis: this was demonstrated by TISELIUS et al. [T 14] in the purification of *turnip yellow mosaic virus*.

7.10. Nucleic Acids

The introduction of polyacrylamide gel as a supporting medium in electrophoresis has offered the possibility of efficient and precise separations of nucleic acids. The greatly improved resolution possible with this method compared to sucrose density gradient centrifugation has been convincingly demonstrated e.g. by WEINBERG et al. [W 6a] using nucleolar RNA. In addition, polyacrylamide gel has the advantage of a low risk of ribonuclease contamination in contrast to agar gels and sugar gradients. The molecular sieving property of the gel allows the electrophoretic fractionation of nucleic acid molecules which differ in size, since they possess a constant charge to size ratio in neutral and alkaline buffer systems. Moreover, the method permits a relatively rapid and simple determination of both sedimentation coefficient and molecular weight (see page 17).

7.10.1. Special Conditions for Optimal Electrophoretic Separations of RNA in Polyacrylamide Gel

During the course of extensive experimental studies on the electrophoretic behavior of RNA in polyacrylamide gels, several investigators found that special conditions should be observed to achieve optimal separations.

The resolving power is highly dependent on the *purity* of acrylamide and Bis used as well as of the RNA applied. Contaminants in the monomers (polymer), catalysts, or in the RNA (proteins) can result in streaking and poor resolution [D 8, L 35]. The purification of the gel monomers and of the RNA therefore merits particular attention (pp. 34 and 41).

To this end, the gels should be prepared from recrystallized acrylamide and Bis, preelectrophoresed and washed to remove any impurity. In addition, WEINBERG et al. [W 6a] shake monomer solutions with hexane at 20° C. These procedures yield gels which can be scanned at 265 nm for RNA detection. Nevertheless a high background of density may be found near the top of the gel, particularly after storage for some days [L 36]. It can be reduced by extensive washing of the gel in buffer (for hours or days) after pre-electrophoresis.

Impure RNA, on the other hand, may aggregate and stick to the top of the gel. RNA that has entered may constrict the top gel region and trail out during electrophoresis. LOENING [L 35] found thorough deproteinization to be essential. Hence the method of RNA preparation is critical. This was also stressed by Davis [D 8], who observed that pattern variations of RNA actually reflect differences in the extraction procedure.

The first 2–3 mm of the top of the gel may be more porous than the rest of the gel owing to dilution of the gel solution during water layering [L 36, P 9]. Similarly, the gel structure of the lower end may be less homogeneous (pp. 16, 37). Therefore, the use of these portions of the gel for the determination of e.g. molecular weights is not recommended. – The polymerization of high-concentration gels (15 %) is better reproducible if the gel tubes are immersed in a constant temperature bath [R 36]. High-molecular weight RNA does not diffuse out of the gel during extended washing, but t-RNA will do so.

RICHARDS and GRATZER [R 36] discussed the choice of discontinuous and continuous buffer systems for optimal RNA separations. They concluded that discontinuous buffer systems may be used if dilute RNA are to be applied. Continuous buffer system are preferred, if concentrated solutions are available. These can be directly put onto the top of the gel; entering RNA, in contrast to protein, is there retarded and concentrated. In selecting discontinuous buffer systems, however, the fact must be taken into account that the mobilities of all but the largest RNAs are, in general, higher than those of most proteins. RICHARDS and GRATZER [R 34, R 36] therefore devised several systems in which the degree of ionization (and hence the effective mobility) of the trailing ion (weak acid) in the separation gel was raised (Table 5). As continuous systems they recommend:

(1) pH 7.4: Tris 1.43 g, 1 N HCl 10 ml, water to 1 l,

(2) pH 5.0: NaOH 4.0 g, glacial acetic acid 8.35 ml, water to 1 l.

7.10.2. Low Molecular Weight RNA and Polynucleotides

The pioneering work of RICHARDS, GRATZER et al. [C 9, R 34–36] has demonstrated that disc electrophoresis can be successfully employed for the separation of molecules of *low molecular weight RNA* (s- and t-RNA, degradation products of m-RNA and r-RNA) and thus for a purity test of RNA preparations (Fig. 82). The authors used discontinuous buffer systems to achieve satisfactory separations of dilute RNA solutions (Table 5, p. 48) and stres-

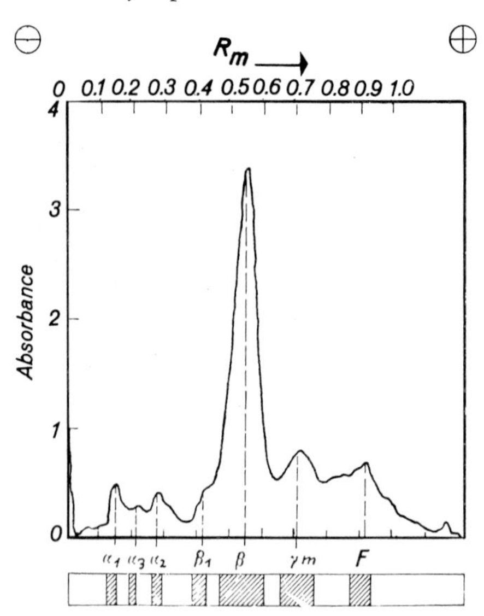

Fig. 82. Disc electrophoresis of soluble ribonucleic acids (s-RNA) with microdensitometer trace. From: RICHARDS et al. [R 34].

R_m = Ratio of migration distance of a component to the migration distance of the anion front (from start to front F). β-Band: Amino acid accepting RNA.

sed the influence of different gel concentrations on the resolution of two components, expressed by the difference of their R_m-values (ΔR_m, Fig. 83). This permits a determination of the gel concentration which will yield an optimal resolution of two components. In addition, they found linear relationships between the extractability of RNA from the gel, the RNA content and the intensity of the RNA stain in the gel as well as between relative

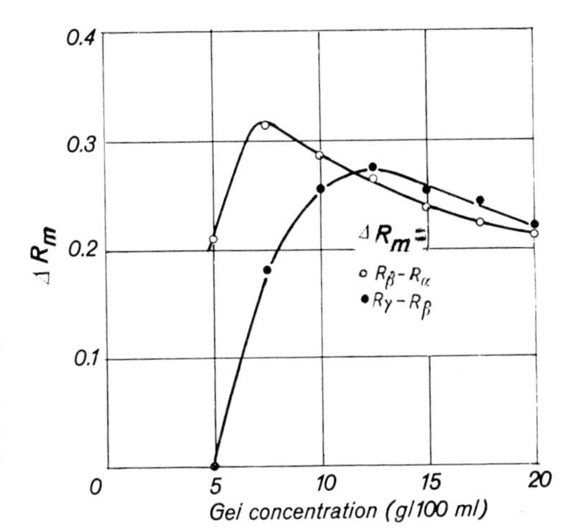

Fig. 83. Relation between the resolution ΔR_m of different RNA components (a_1, β, γ_m; see Fig. 82) and the gel concentration (% acrylamide, w/v). From: RICHARDS et al. [R 34].

electrophoretic mobility and sedimentation coefficient (see page 17). Fig. 11 (p. 18) shows that the R_m-value permits an estimate of the sedimentation coefficient.

To prepare suitable gel solutions, a mixture of acrylamide and Bis, in a ratio calculated according to eqn. (3) on page 3, is added to the large and small pore solutions (Table 5) to the desired concentration depending on the MW range of the RNA to be resolved. For the fractionation of s-RNA, RICHARDS and GRATZER [R 36] recommend the following mixture (e.g. pH 8.9 or 9.2 system, Table 5). Separation (small-pore) gel: 9.5 g acrylamide + 0.5 g Bis + small-pore buffer to 100 ml + 0,1 ml TEMED. Spacer (large pore) gel: 4.75 g acrylamide + 0.25 g Bis + large-pore buffer to 100 ml + 0.1 ml TEMED. Each solution is degassed in a syringe and rapidly mixed with one tenth of the volume of an 1 % ammonium persulfate solution. Acrylamide solutions of 15 % are precooled before mixing to reduce the polymerization rate.

About 100–200 µg RNA are required as an optimum load for 5 mm diameter gels. But less amount may be detected by staining (page 79). 5 % Sucrose and a few µl of a 0.001 % bromophenolblue solution are added to the sample solution which is put on top of the gel and carefully overlayered with electrode buffer. Alternatively, the sample solution is soaked into one or several small filter paper disks which are placed on the gel and covered with a slurry of sand. For the use of continuous buffer systems, it is essential to keep the ionic strength several times lower than that of the gel (page 25). Electrophoresis at 5 mA/gel, 40 min.

Using 5 % separation gels with the pH 8.9 buffer system (Table 5), however without spacer gels, McPHIE et al. [M 40] and GOULD et al. [G 26–28] found distinct RNA fragments after degradation of *r-RNA* with T_1 and pancreatic ribonuclease. The patterns obtained were characteristic for the RNA species studied (*RNA finger printing analysis*). HINDLEY [H 22] used "finger printing" patterns of highly radioactive t-RNA digests for sequencing. GEROCH et al. [G 5] separated E. coli 5S-RNA from 4S-RNA in 10 % separation gels with a continuous buffer system (0.05 M Tris acetate pH 7.0).

Even smaller ribosomal and viral RNA fragments (MW 10^4 down to small *oligonucleotides* of N = 10-20) have been separated on analytical and preparative gels (T = 15 %, C = 1 %) by GOULD et al. [G 29] using discontinuous (pH 8.9 buffer, Table 5) and continuous buffer systems (5 mM acetate buffer pH 5). BIRNBOIM and GLICKMAN [B 30] fractionated oligonucleotides (tetra- to nonanucleotides) of equal chain length (isopliths) but different net charge in 8 and 4 % gels using acid buffer systems (0. 1 M formic acid, pH 2.3; 0.05 M sodium citrate, pH 2.9–3.4; 0.1 and 0.2 M acetic acid).

7.10.3. High Molecular Weight RNA

The thorough studies by LOENING [L 35–37] and others (see page 181), using continuous buffer systems, as well as by GROSSBACH and WEINSTEIN [G 43], employing discontinuous systems, have demonstrated that even *high molecular weight RNAs* can be fractionated with high resolving capacity in large pore polyacrylamide gels.

GROSSBACH and WEINSTEIN [G 43] obtained sharp fractionations not only of soluble but also of high molecular weight RNA (r- and presumably also m-RNA) from Chironomus larvae and Drosophila flies with *discontinuous gradient gels* of 7.18 and 2.5 % total monomer T (Fig. 84).

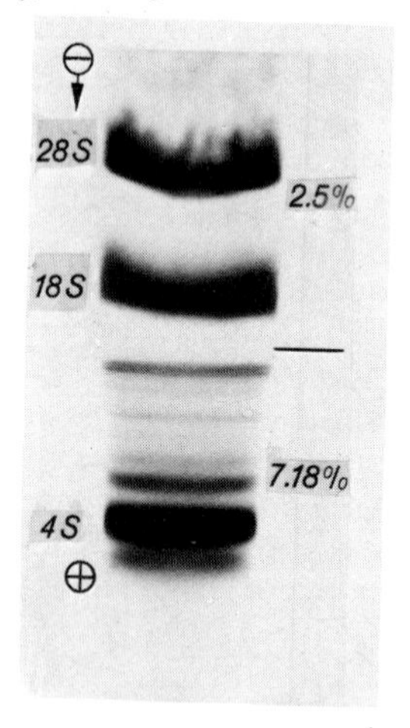

Fig. 84. Disc electrophoretic fractionation of RNA from the larvae of Chironomus tentans in a discontinuous gradient gel containing 7.18 and 2.5% total monomer. Gallocyanine-chrome alum stain (p. 79). From: GROSSBACH and WEINSTEIN [G 43].

The buffers of gel system No. 6 (Table 4) are used except that solution No. 16 is replaced by No. 16a: 1 N HCl 48 ml, Tris 4.95 g, TEMED 0.46 ml, water to 100 ml, pH 5.5. A "large-pore" separation gel (1.5 cm length, T = 2.5, C = 5 %) is polymerized over a "small-pore" separation gel (containing 7 % acrylamide, 0.186 % Bis) of several cm length. In place of the usual spacer (stacking) gel the combined separation gel is overlaid with 0.2 ml of spacer gel buffer containing 20 % sucrose. The solution of 5–100 μg RNA in 10–200 μl 0.01 M acetate buffer pH 5.1 (with 10^{-4} M $MgCl_2$) is mixed with an equal volume of double strength spacer gel buffer and applied on to the top of the standard gel column (5 mm diameter). Electrophoresis for 1 hr. at 1 mA/ gel and 4° C, then 1–3 hrs. at 2 mA/gel. RNA staining, see page 79. Direct densitometry is used for quantitation (orange filter, Ilford 607). A linear relation exists between the absorbance of the dye complex and the quantity of RNA up to 1.7 mg RNA/ml gel. To suppress artifacts of direct densitometry which result from the deflection of the light beam at the boundary between gel layers, the authors prefer a densitometric evaluation of photographs (Polaroid transparent film 46–L or 55 P/L with Kodak Wratten Filter No. 25 and diffuse, transmitted light).

The authors found that (1) 28 S RNA migrate into 2.5 % gels, barely enter 2.75 % gels and is excluded from 3.3 % gels, (2) 18 S RNA is excluded from 4 % gels, and (3) 4 S RNA travel with the salt front in less than 7 % gels while in 7 or 10 % gels it moves slower than the front. Minor RNA components, not detectable by sucrose gradients, are observed between the 10 S and 4 S RNAs.

The pioneering work by Loening [L 35–37] has indicated that rather long (up to 10 cm) large pore gels containing continuous buffer systems and SDS are also highly capable to fractionate high and low molecular weight RNAs from diverse sources, for example, BMV and TMV virus, E. coli, pea seedling root tips, Drosophila flies, Xenopus tadpoles, reticulocytes, HeLa cells (Fig. 85). Thus, 2.2–2.6 % gels resolved the ribosomal RNA components, the 45 S precursor, t-RNA and minor components so far not seen in sucrose gradients. Gels prepared as described on page 41, could be easily and rapidly scanned at 265 nm (background absorption less than 0.3) to detect RNA with a limit of about 0.05 μg RNA in a zone of 1 mm width (gel diameter 6.3 mm). According to Lewicki and Sinskey [L 21a], the precision of RNA separation in these gels is high and separation of two RNA species differing by 1 S unit is easily obtained.

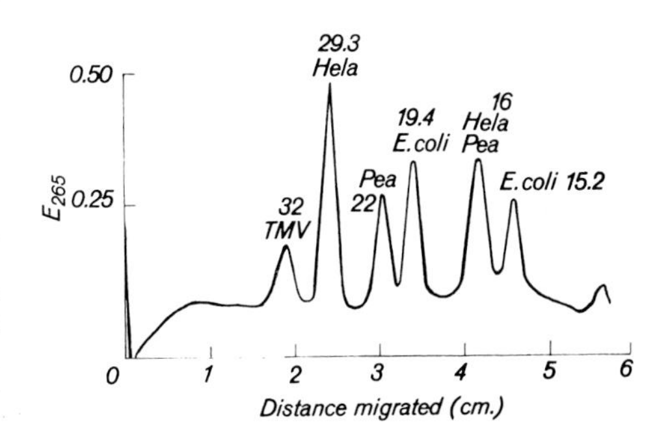

Fig. 85. Electrophoresis of a mixture of distinct RNA species. From: Loening [L 36].

A 20 μl sample of a mixture containing 50–80 μg of each of HeLa-cell r-RNA, pea r-RNA, E. coli r-RNA and TMV-RNA in 200 μl E buffer (see text, with SDS and sucrose) was layered on a 2% gel and electrophoresed for 2.5 hrs. Then the gel was immediately scanned at 265 nm. The numbers for each RNA represent S-values.

Stock acrylamide solutions:

Recrystalized acrylamide 15 g
Recrystalized Bis 0.75 g for 2-5% gels, 0.375 g for > 5% gels
Water to a total volume of 100 ml

Stock 5 x concentrated buffer solutions:

	Acetate buffer	*Phosphate buffer*
Tris	24.2 g (40 mM)	21.8 g (36 mM)
Sodium acetate anhydr.	8.2 g (20 mM)	–
$NaH_2PO_4 \cdot 2H_2O$	–	23.4 g (30 mM)
$Na_2EDTA \cdot 2H_2O$	1.85 g (1 mM)	1.85 g (1 mM)
Water to a total vol. of	1 l	1 l (pH 7.7)
Acetic acid to pH	7.8	–

Final running concentrations are given in parentheses.

To prepare gel solutions of desired concentrations five volume parts of stock acrylamide solution are mixed with the following volume parts of buffer and water:

Gel conc. %	2.0	2.2	2.4	2.5	2.6	3.0	5.0	7.5
Buffer	7.5	6.8	6.25	6.0	5.8	5.0	3.0	2.0
Water	24.7	22.0	19.7	18.7	17.8	14.7	6.7	2.7

Following degasing under vacuum for 15 sec. 15 μl of TEMED and 0.25 ml of freshly dissolved 10% aqueous ammonium persulfate are added. A five times dilution of stock buffer is used as electrode (E) buffer to which 0.2% SDS may be added. Preelectrophoresis for 30-60 min. at 8 V/cm gel length (5 mA/6 mm gel diameter) and 25°C is recommended. Then the RNA sample (up to 2 mg/ml E buffer in addition containing 5% sucrose) is applied and electrophoresis continued for 30 min. to 4 hrs. After the run, the gels are blown out of the Plexiglas (Lucite) tubes with a rubber teat and scanned at 265 nm immediately. They are best picked up by sucking into tubes of same diameter.

LOENING's systems were further exploited by BISHOPS et al. [B 31–32] and MILLS et al. [M 50] for the separation of 10 virus nucleic acids and 4 E. coli RNAs and by WEINBERG et al. [W 6a] for the fractionation of HeLa nucleolar r-RNAs. BISHOPS et al. found [B 31] that virus RNAs do not suffer depolymerization or loss of biological activity during electrophoresis. Similar good resolutions of high molecular weight RNAs from rat liver, kidney and brain were obtained by PEACOCK and DINGMAN [P 8] using a continuous Tris – borate – EDTA buffer system in a vertical gel slab.

Gel and electrode buffer (pH 8.3): Tris 10.8 g, Na_2EDTA 0.93 g, boric acid 5.5 g water to 1 l. Gel concentrations: T = 3.5 and 10 %, C = 5 % (constant), 0.44 % DMPN and 0.1 % ammonium persulfate. Electrophoresis of 3 mm thick gel slabs at 200 V (10 V/cm) and 50 mA for 1.5–4 hrs. under cooling (0° C). RNA staining see page 79.

However, low concentration gels (T < 3.5 %) of dimensions exceeding those of the standard gel cylinders are too soft to be handled without risk of damaging. The early report of URIEL [U 1–2] had indicated that the incorporation of *agarose* in polyacrylamide gels provides gels of excellent mechanical properties (see page 66). DINGMAN and PEACOCK [D 20, P 9] then explored the applicability of such *composite* gels and were able to prepare gels of very low concentration (T = 2 %, C = 5 %, 0.5 % agarose) which retain sufficient strength to allow easy managing. Interestingly, the *agarose gel "corset"* pervading the polyacrylamide gel matrix exerts, under the conditions employed, no detectable effects on the sieving properties of polyacrylamide gels as known from "pure" gels. Composite gels were successfully used by DINGMAN and PEACOCK [D 20] for high-resolution analysis of RNAs in the size from 10^4 up to 10^8 daltons (nuclear and cytoplasmic rat liver RNAs).

Agarose is dissolved in water with heating, cooled to 40° C, and mixed with a solution also at 40° C, containing acrylamide, Bis, dimethylaminopropionitrile and buffer. Ammoniumpersulfate solution (not warmed) is added and the solution poured into the slab mold previously equilibrated at 20° C. The temperature is kept at 20° C for about 1 hr. until the solution is polymerized. A low concentration of the catalyst (0.05 %) ensures the gelation of agarose prior to that of acrylamide.

Similarly, WATANABE et al. [W 4] fractionated *reovirus RNAs*, PONS and HIRST [P 25] *influenza virus RNAs* and HOTHAM-IGLEWSKI et al. [H 41] *r-RNAs of bacteriophage R* 17 infected E. coli in 2.2–2.4 % gels containing 1 % agarose. Even less than 1 ng of RNAs microisolated from subcellular fractions of Chironomus salivary gland cells could be separated in horizontal microgel slides containing 2.7 % acrylamide (C = 5 %) and 1 % agarose [R 42].

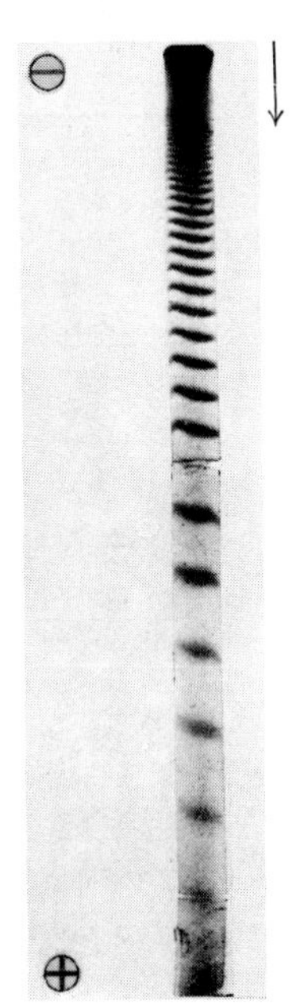

Fig. 86. High resolution analytical disc electrophoresis of a *d AT digest* (by pancreatic DNAse) in a gel containing 22% acrylamide and 3% Bis. From: ELSON and JOVIN [E 4].

Gel length: 140, diameter: 5 mm. Buffer systems see Table 8. Electrophoresis at 1–4 mA/gel. Toluidine blue 0 staining (see page 79).

7.10.4. DNA and Polydeoxynucleotides

Few attempts have been made to electrophorese *DNA* in polyacrylamide gels. LOENING [L 35] found that high molecular weight DNA enters 2.4 % gels as a single sharp peak with a low relative mobility. DNA which has been ultrasonically degraded to a molecular weight of about 5×10^5, moves as a very diffuse band with a mobility comparable to r-RNA, whereas DNAse degraded DNA travels ahead of 4S RNA. According to PEACOCK and DINGMAN [P 8], DNA as contaminant in RNA preparations from rat kidney and brain cytoplasm migrates in 3. 5 % gels to the position of the slower 18 S RNA band.

Disc electrophoresis appears to be superior to ion exchange chromatography in resolving *oligodeoxynucleotides* of high degree of polymerization. ELSON and JOVIN [E 4] were able to fractionate mixtures of the oligomers $d(AT)_n$, produced by enzymic digestion of dAT, on analytical gels over the range from n = 3 to about n = 30 (Fig. 86) and on preparative gels over a range from n = 3 to about n = 23.

The separation gels contain 19.5–22 % acrylamide and 1–3 % Bis and are 10–15 cm long. On problems arising with the use of such highly concentrated gels see page 56.

8. APPLICATIONS IN FOOD TECHNOLOGY AND FORENSIC ANALYSIS

Since disc electrophoresis represents an excellent tool to monitor protein transformations and deteriorations which frequently occur during storage of foods, the method is of importance in food technology. In addition, the technique permits a sensitive test of *food adulteration* because of the specifity of the protein patterns obtained from the particular sources. Moreover the need of only minute sample amounts makes it most suitable for *forensic analysis*. Thus, adulterations of marketed fish were detected [M 5, P 7] (see page 163). The question was tested how long one should store and cook ham [C 46] or beef [S 8] or age the chicken breast [M 1a]. During 8-week storage at –4° C, the total protein content of bovine muscle drops nearly to 50 % and the disc pattern of both sarcoplasmic proteins and actomyosin changes considerably [A 45]. Flax cells stored at –50° C for 14 days show no qualitative differences in banding pattern, but quantitative changes in specific fractions [Q 1].

For *dairy* scientists the method is of increasing significance since it allows to answer questions such as whether hydrogenperoxide as bactericide in milk does damage its protein structure [G 37] or whether reconstitution of milk powder does not change the film formed at air – water interface [L 10]. A 15 days storage of sterile unheated skimmilk at + 4° C does not change the whey protein pattern, but storage at + 30° C will do so [M 49].

Disc electrophoresis allows, for example, to identify mottled and unmottled *wheat* [B 5], different wheat seeds [J 9–11] (p. 170), *potato* varieties [M 1, S 41a], to determine the type and amount of durum and soft wheat used in various pastas [R 29–30] (Fig. 87) and to quantitate the amount of *egg* used in the manufacture of noodles [S 42–43] (Fig. 88). SILANO et al. [S 42] developed a

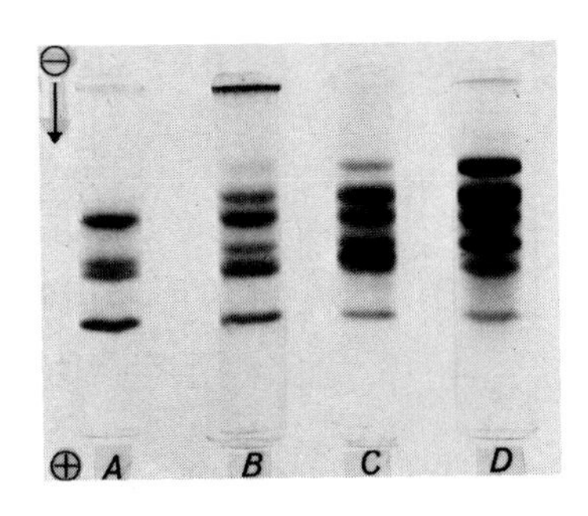

Fig. 87: Protein patterns of pure durum wheat (A), pure soft wheat (B), and of mixtures of durum wheat with 10 % (C) and 30 % (D) soft wheat. From: RESMINI [R 29].

Proteins from semolina or pastas were extracted with magnesium and ammonium sulfate solutions and precipitated by addition of highly concentrated ammonium sulfate solutions. Electrophoresis with gel system No. 1 at 3.7 mA/gel, 60–200 V, 80 min. Amido Black staining.

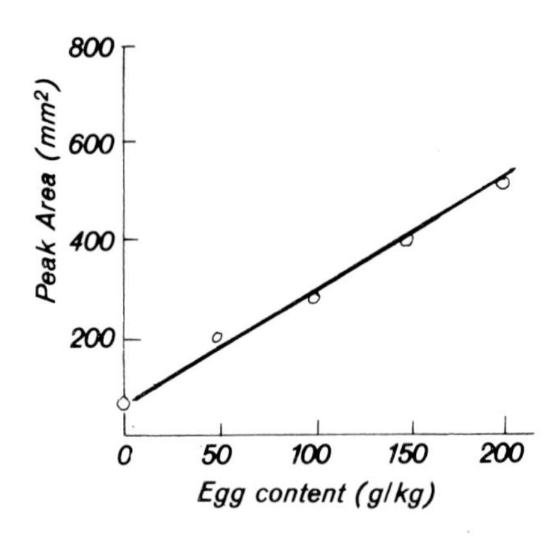

Fig. 88: Relationship between egg content in noodles and area under the peak of an egg white (albumin) protein band found in disc electropherograms of noodle extracts. From: SILANO et al. [S 43].

quantitative test for *barley* adulteration in wheat flours and macaronis. Official bodies (e.g. US Food and Drug Administration) now utilize disc electrophoresis to provide evidence of illegal adulteration of food products.

RESMINI's technique [R 29–30] to identify the addition of soft wheat (Triticum aestivum) in durum wheat (T. durum) products, i.e. kernel wheat, semolina or pasta, is of apparent significance since Italian and French laws allow manufacture of such cereal food only from durum wheat.

The method by SILANO et al. [S 43] to determine the egg content of noodles represents an important supplement to the official AOAC sterol method in that it gives a true analysis of the composition of these commercial products. The AOAC method is based on measurement of sterols which are present only in the yolk. Thus, the protein method combined with sterol determination can detect a fraud which is rather frequent, i.e. the addition of a lesser amount of egg white with the prescribed amount of yolk.

References

A 1 ABADI, D. M.: Clin. Chem. *15*, 35 (1969).
A 2 ABE, M., KRAMER, S. P., and SELIGMAN, A. M.: J. Histochem. Cytochem. *12*, 364 (1964).
A 3 ABIKO, Y., IWAMOTO, M., and SHIMIZU, M.: J. Biochem. *64*, 743 (1968).
A 4 ACKERS, G. K.: Biochemistry *3*, 723 (1964).
A 5 ADAMS, J. B., and WACKER, A.: Clin. Chim. Acta *21*, 155 (1968).
A 6 ADAMS-MAYNE, M. E., and JIRGENSONS, B.: Arch. Biochem. Biophys. *113*, 575 (1966).
A 7 ADELMANN, B., MARQUARDT, H., and KÜHN, K.: Biochem. Z. *346*, 282 (1966).
A 8 AFFRONTI, L. F., PARLETT, R. C., and CORNESKY, R. A.: Amer. Rev. Respir. Dis. *91*, 1 (1965).
A 9 AKAZAWA, I., SAIO, K., and SUGIYANA, N.: Biochem. Biophys. Res. Commun. *20*, 114 (1965).
A 10 AKHTAR, A., HANSEN, A., and KÄRCHER, K. H.: Z. klin. Chem. u. klin. Biochem. *6*, 334 (1968).
A 11 AKROYD, P.: Anal. Biochem. *19*, 399 (1967).
A 12 AKROYD, P.: Chromatogr. and Electr. Techniques, Vol. II. I. Smith editor, W. Heinemann Med. Books, London, 1968, p. 399.
A 13 ALKJAERSIG, N.: Biochem. J. *93*, 171 (1964).
A 14 ALLEN, J. M.: Annals N. Y. Acad. Sci. *94*, 937 (1961).
A 15 ALLEN, J. M., and BEARD, M. E.: Science *149*, 1507 (1965).
A 16 ALLEN, J. M., and GOCKERMAN, J.: Annals N. Y. Acad. Sci. *121*, 616 (1964).
A 17 ALLEN, J. M., and HYNCIK, G. J.: J. Histochem. Cytochem. *11*, 169 (1963).
A 18 ALLEN, R. C.: ORTEC Life Sci. *1*, No. 1 (1969).
A 19 ALLEN, R. C., and JAMIESON, G. R.: Anal. Biochem. *16*, 450 (1966).
A 20 ALLEN, R. C., and MOORE, D. J.: Anal. Biochem. *16*, 457 (1966).
A 21 ALLEN, R. C., and MOORE, D. J.: Endocrin. *78*, 655 (1966).
A 22 ALLEN, R. C., MOORE, D. J., and DILWORTH, R. H.: J. Histochem. Cytochem. *17*, 189 (1969).
A 22a ALLEN, R. C.: Personal communication (1970).
A 23 ALLEN, R. C., POPP, R. A., and MOORE, D. J.: J. Histochem. Cytochem. *13*, 249 (1965).
A 24 ALLEN, S. L.: Brookhaven Symp. in Biol. *18*, 27 (1965).
A 24a ALLRED, R. J., and KEUTEL, H. J.: J. Lab. Clin. Med. *71*, 179 (1968).
A 24b ALPERS, D. H., and GLICKMAN, R.: Anal. Biochem. *35*, 314 (1970).
A 25 ALTSCHUL, A. M., EVANS, W. J., CARNEY, W. B., MCCOURTNEY, E. J., and BROWN, H. D.: Life Sci. *3*, 611 (1964).
A 26 AMARAL, L.: Canalco Abstract No. 1170 (1969).
A 27 AMELUNXEN, R. E.: Biochim. Biophys. Acta *122*, 175 (1966).
A 28 AMERICAN CYANAMID CO.: Chemistry of Acrylamide, a technical bulletin (1965).
A 29 ANDERSON, P. J., SONG, S. K., and CHRISTOFF, N.: Proc. 4th Intern. Congr. Neuropath. Vol. I, p. 75 (1962).
A 29a ANKER, H. S.: FEBS Letters *7*, 293 (1970).
A 29b ANTOINE, B.: Science 138, 977 (1962)
A 30 ANTWEILER, H. J., in: Methoden der org. Chemie (Houben-Weyl), Thieme, Stuttgart (1955), Vol. III, Part 2, p. 214.
A 31 AOKI, N., and KAULLA, K. N.: Proc. Soc. Exp. Biol. *130*, 101 (1969).
A 32 AOYAGI, R., ASAGAMI, Y., and MOGI, G.: Seibutsu-Butsuri-Kagaku (Japan) *11*, 69 (1965).

A 33 Apirion, D.: J. Mol. Biol. *16,* 285 (1966).
A 34 Apostolopoulos, A. X.: J. Dental Res. *43,* 766 (1964).
A 35 Appel, S. H., Alpers, D. H., and Tomkins, G. M.: J. Mol. Biol. *11,* 12 (1965).
A 36 Arai, K., and Watanabe, S.: J. Biol. Chem. *243,* 5670 (1968).
A 37 Aronson, J. N., and Borris, D. P.: Anal. Biochem. *18,* 27 (1967).
A 38 Aronson, N. N., and Davidson, E. A.: J. Biol. Chem. *242,* 437 (1967).
A 39 Asagami, Y.: Yamaguchi Igaku (Japan) *14,* 187 (1965).
A 40 Asagami, Y., Mogi, G., Aoyagi, R., and Inadomi, H.: Seibutsu-Butsuri-Kagaku (Japan) *11,* 68 (1965).
A 41 Asagami, Y., Mogi, G., Sasaki, I., Aoyagi, R., and Nakamura, S.: Seibutsu-Butsuri-Kagaku (Japan) *10,* 98 (1964).
A 42 Asagami, Y., Nakamura, K., Mogi, G., and Nakamura, S.: Shinkei Kenkyu no Shinpo (Japan) *8,* 653 (1964).
A 43 Asakura, S., Eguchi, C., and Iino, T.: J. Mol. Biol. *16,* 302 (1966).
A 44 Atherton, L., and Hershberger, W.: Canalco Newsletter No. 8, *3* (1966).
A 45 Awad, A., Powrie, W. D., and Fennema, O.: J. Food Sci. *33,* 227 (1968).
A 46 Awdeh, J. L., Williamson, A. R., and Askonas, B. A.: Nature *219,* 66 (1968).
A 47 Axline, S. G.: J. Exper. Med. *128,* 1031 (1968).

B 1 Bacchus, H., Kennedy, G. R., and Blackwall, J.: Cancer *20,* 1654 (1970).
B 2 Baglioni, C., Cioli, D., Gorini, G., Ruffilli, A., and Alescio-Zonta, L.: Cold Spring Harb. Symp. *32,* 147 (1967).
B 3 Baldwin, A. N., and Berg, P.: J. Biol. Chem. *241,* 831 (1966).
B 4 Barber, J. T., and Steward, F. C.: Devel. Biol. *17,* 326 (1968).
B 5 Barber, J. T., Wood, H. L., and Steward, F. C.: Canad. J. Botan. *45,* (1967).
B 6 Barka, T.: J. Histochem. Cytochem. *9,* 542 (1961).
B 7 Barka, T.: J. Histochem. Cytochem. *9,* 564 (1961).
B 8 Barka, T., and Anderson, P. J.: Histochemistry, Harper and Row, New York (1963), p. 313.
B 9 Barker, J. E.: Develop. Biol. *18,* 14 (1968).
B 10 Basch, R. S.: Anal. Biochem. *26,* 184 (1968).
B 11 Bauer, G. E., Lindall, A. W., Dixit, P. K., Lester, G., and Lazarow, A.: J. Cell Biol. *28,* 413 (1966).
B 12 Bauer, R. D., and Johanson, R.: Biochim. Biophys. Acta *119,* 418 (1966).
B 13 Bauer, R. D., and Thomasson, W. A.: J. Chromatog. *22,* 496 (1966).
B 14 Bednarik, T.: Clin. Chim. Acta *15,* 172 (1967).
B 15 Bednarik, T.: Clin. Chim. Acta *17,* 132 (1967).
B 16 Beers, R. F., and Armilei, G.: Nature *208,* 466 (1965).
B 17 Benjamin, W., and Gellhorn, A.: Proc. Nat. Acad. Sci. *59,* 262 (1968).
B 18 Bennick, A.: Anal. Biochem. *26,* 453 (1968).
B 19 Bensusan, H. B.: Arch. Biochem. Biophys. *115,* 77 (1966).
B 20 Benton, A. W., and Patton, R. L.: J. Insect. Physiol. *11,* 1359 (1965).
B 21 Berkman, P. M., Pastewka, J. V., and Peacock, A. C.: Biochim. Biophys. Acta *181,* 159 (1969).
B 22 Berry, S. J., Krishnakumaran, A., and Schneiderman, H. A.: Science *146,* 938 (1964).
B 23 Besemer, J., and Clauss, H.: Z. Naturforschg. *23b,* 707 (1968).
B 24 Biedermann, K., and Frieser, H.: Rö.-Bl. *20,* 157 (1967).
B 25 Biel, H., and Zwisler, O.: Beringwerk-Mitteilungen *46,* 141 (1966).
B 26 Billiar, R. D., Brungard, J. C., and Villee, C. A.: Science *146,* 1464 (1964).
B 27 Birk, Y., and Li, C. H.: J. Biol. Chem. *239,* 1048 (1964).

B 28 Birks, J. B.: Theory and Practice of Scintillation Counting, Macmillan
B 29 Birnboim, H. C.: Anal. Biochem. *29,* 498 (1969).
B 30 Birnboim, H. C., and Glickman, J.: J. Chromatog. *44,* 581 (1969).
B 31 Bishop, D. H. L., Claybrook, J. R., and Spiegelman, S.: J. Mol. Biol. *26,* 373 (1967).
B 32 Bishop, D. H. L., Claybrook, J. R., Pace, N. R., and Spiegelman, S.: Proc. Nat. Acad. Sci. *57,* 1474 (1967).
B 33 Bishop, S., and Grisolia, S.: Biochim. Biophys. Acta *118,* 211 (1966).
B 34 Björk, I.: Exper. Eye Res. *7,* 129 (1968).
B 35 Blattner, D. P.: Anal. Biochem. *27,* 73 (1969).
B 35a Blattner, D. P., and Reithel, F. J.: J. Chromatog. *46,* 286 (1970).
B 36 Bloemendal, H.: Chemisch Weekblad *59,* 281 (1963).
B 37 Bloemendal, H., Bont, W. S., Benedetti, E. L., and Wisse, J. H.: Exper. Eye Res. *4,* 319 (1965).
B 38 Bloemendal, H., Bont, W. S., Jongkind, J. F., and Wisse, J. H.: Exper. Eye Res. *1,* 300 (1962).
B 39 Bloemendal, H., Jongkind, J. F., and Wisse, J. H.: Chemisch Weekblad *58,* 501 (1962).
B 40 Boedtker, H.: J. Mol. Biol. *2,* 171 (1960).
B 41 Boettcher, B., and De La Lande, F.: Anal. Biochem. *28,* 510 (1969).
B 42 Bonavida, B., Sapse, A. T., and Servarz, E. E.: Nature *221,* 375 (1969).
B 43 Bonavita, V., and Scardi, V.: J. Chromatog. *1,* 285 (1958).
B 44 Bonner, J., Chalkley, G. R., Dahmus, M., Fambrough, D., Fujimura, F., Huang, R. C. C., Huberman, J., Jensen, R., Marushige, K., Ohlenbusch, H., Olivera, B., and Widholm, J.: Methods in Enzymology, Vol. 12, Nucleic Acids, Part B, p. 3, Grossman, L., edit., Academic Press (1968).
B 45 Bont, W. S., Geels, J., and Rezelman, G.: Anal. Biochem. *27,* 99 (1969).
B 46 Bornstein, P., Martin, G. R., and Piez, K. A.: Science *144,* 1220 (1964).
B 47 Bornstein, P., and Piez, K. A.: Science *148,* 1353 (1965).
B 48 Borris, D. P., and Aronson, J. N.: Anal. Biochem. *22,* 546 (1968).
B 48a Borris, D. P., and Aronson, J. N.: Anal. Biochem. *32,* 273 (1969).
B 49 Bowden, J. A., and Connelly, J. L.: J. Biol. Chem. *243,* 3526 (1968).
B 50 Boyd, J. B., and Mitchell, H. K.: Anal. Biochem. *13,* 28 (1965).
B 51 Boyd, J. B., and Mitchell, H. K.: Anal. Biochem. *14,* 441 (1966).
B 52 Boyd, J. B., and Mitchell, H. K.: Arch. Biochem. Biophys. *117,* 310 (1966).
B 53 Brackenridge, C. J., and Bachelard, H. S.: J. Chromatog. *41,* 242 (1969).
B 54 Brattsten, I.: Archiv för Kemi *8,* 205 (1955).
B 55 Bray, G.: Anal. Biochem. *1,* 279 (1960).
B 56 Brewer, J. M.: Science *156,* 256 (1967).
B 57 Bron, C., Blanc, B., and Isliker, H.: Biochim. Biophys. Acta *154,* 61 (1968).
B 58 Broome, J.: Nature *199,* 179 (1963).
B 59 Brownstone, A. D.: Anal. Biochem. *27,* 25 (1969).
B 60 Burger, W. C., and Wardrip, W. O.: Anal. Biochem. *13,* 580 (1965).
B 61 Burger, M. M., and Goldberg, A. R.: Proc. Nat. Acad. Sci. *57,* 359 (1967).
B 62 Burns, D. A., and Pollak, O. J.: J. Chromatog. *11,* 559 (1963).
B 62a Burr, I. M., Balant, L., Stauffacher, W., and Renold, A. E.: Anal. Biochem. *35,* 82 (1970).
B 63 Burrows, S.: Clin. Chem. *11,* 1068 (1965).
B 64 Burtin P., in: Immuno-elektrophoretische Analyse, P. Grabar and P. Burtin, edit., Elsevier Publ. Co., Amsterdam, German translation (1964), p. 103 f.
B 65 Bush, E. T.: Anal. Chem. *35,* 1024 (1963).

C 1 Cahn, R. D.: J. Cell Biol. *27,* 17A (1965).
C 2 Cain, D. F., and Pitney, R. E.: Anal. Biochem. *22,* 11 (1968).
C 3 Caldwell, R. C., and Pigman, W.: J. Dental Res. *43,* 765 (1964).
C 4 Caldwell, R. C., and Pigman, W.: Arch. Biochem. Biophys. *110,* 91 (1965).
C 5 Calendar, R., and Berg, P.: Biochemistry *5,* 1681 (1966).
C 6 Campbell, B. J., Lin, Y. C., Davis, R. Y., and Ballew, E.: Biochim. Biophys. Acta *118,* 371 (1966).
C 7 Canalco (Firm 5, p.): Prep. Disc. Instruction Manual (1967).
C 8 Cann, J. R.: Biochemistry *5,* 1108 (1966).
C 9 Cannon, M., and Richards, E. G.: Biochem. J. *103,* 23C (1967).
C 10 Carpenter, E. L., and Davis, H. S.: J. appl. Chem. *7,* 671 (1957).
C 11 Carrel, S., Gerber, H., and Barandun, S.: Nature *221,* 385 (1969).
C 11a Carrel, S., Theilkaes, L., Skvaril, S., and Barandun, S.: J. Chromatog. *45,* 483 (1969).
C 12 Catsimpoolas, N.: Anal. Biochem. *19,* 592 (1967).
C 13 Catsimpoolas, N.: Anal. Biochem. *26,* 480 (1968).
C 14 Catsimpoolas, N.: Clin. Chim. Acta *23,* 237 (1969).
C 15 Catsimpoolas, N.: Biochim. Biophys. Acta *175,* 214 (1969).
C 16 Catsimpoolas, N.: Science Tools *16,* 1 (1969).
C 17 Catsimpoolas, N., Ekenstam, C., Rogers, D. A., and Meyer, E. W.: Biochim. Biophys. Acta *168,* 122 (1969).
C 17a Catsimpoolas, N., Campbell, T. G., and Meyer, E. W.: Plant Physiol. *43,* 799 (1968).
C 17b Catsimpoolas, N., and Meyer, E. W.: Arch. Biochem. Biophys. *125,* 742 (1968).
C 17c Catsimpoolas, N., and Meyer, E. W.: J. Agr. Food. Chem. *16,* 128 (1968).
C 18 Cawley, L. P.: J. Kansas Med. Soc. *64,* 470 (1963).
C 19 Cawley, L. P., Eberhardt, L., Goodwin, W., Schneider, D., and Harrough, J.: Transfus. *4,* 315 (1964).
C 20 Cawley, L. P., Eberhardt, L., Schneider, D., and Goodwin, W.: Transfus. *5,* 375 (1965).
C 21 Cawley, L. P., Sanders, J. A., and Eberhardt, L.: Amer. J. of Clin. Path. *40,* 429 (1963).
C 22 Cayle, T., Saletan, L. T., and Lopez-Ramos, B.: Wallenstein Lab. Commun. *27,* No. 93/94, 87 (1964).
C 23 Celis, J. E., and Conway, T. W.: Proc. Nat. Acad. Sci. *59,* 923 (1968).
C 24 Center, E. M., Hunter, R. L., and Dodge, A. H.: Genetics *55,* 349 (1967).
C 25 Chader, G. J., and Westphal, U.: Biochemistry *7,* 4272 (1968).
C 26 Chaet, A. B., and Bau, D.: Biol. Bull. *123,* 490 (1962).
C 27 Chaiet, L., and Wolf, F. J.: Arch. Biochem. Biophys. *106,* 1 (1964).
C 28 Chalkley, G. R., and Maurer, H. R.: Proc. Nat. Acad. Sci. *54,* 498 (1965).
C 28a Chalvardjian, A.: Clin. Chim. Acta *26,* 174 (1969).
C 29 Chan, W., Morse, D. E., and Horecker, B. L.: Proc. Nat. Acad. Sci. *57,* 1013 (1967).
C 30 Chance, R. E., Ellis, R. M., and Bromer, W. W.: Science *161,* 165 (1968).
C 31 Chang, L. O., Srb, A. M., and Steward, F. C.: Nature *193,* 756 (1962).
C 32 Chen, P. S.: Experientia *23,* 483 (1967).
C 33 Chino, H., and Gilbert, L. I.: Science *143,* 359 (1964).
C 34 Choules, G. L., and Zimm, B. H.: Anal. Biochem. *13,* 336 (1965).
C 35 Chrambach, A.: Anal. Biochem. *15,* 544 (1966).
C 36 Chrambach, A., Reisfeld, R. A., Wyckoff, M., and Zaccari, J.: Anal. Biochem. *20,* 150 (1967).
C 37 Christoff, N., Anderson, P. J., Slotwiner, P., and Song, S. K.: Annals N. Y. Acad. Sci. *135,* 150 (1966).

C 38 Cintron-Rivera, A. A., Conde del Pino, E., Maldonado, N., and Seneriz, R.: Ann. Mtg. Puerto Rico Med. Soc., Nov. 1964; Canalco Abstract No. 146 B (1965).

C 39 Clare, B., Weber, D. J., and Stahmann, M. A.: Science *153,* 62 (1966).

C 40 Clark, C. C., and Veis, A.: Biochim. Biophys. Acta *154,* 175 (1968).

C 41 Clarke, J. T.: Annals N. Y. Acad. Sci. *121,* 428 (1964).

C 42 Clarke, K., Gray, G. W., and Reaveley, D. A.: Biochem. J. *105,* 755 (1967).

C 42a Clausen, J., in: Laboratory Techniques in Biochem. and Mol. Biol., Vol. 1, part III, Work, T. S., and Work, E., editors, North-Holland Publ. Co., Amsterdam-London (1969), p. 397.

C 43 Clements, R. L.: Anal. Biochem. *13,* 390 (1965).

C 44 Coates, M. L., and Twitty, V. C.: Proc. Nat. Acad. Sci. *58,* 173 (1967).

C 45 Coddington, A., Fincham, J. R. S., and Sundaram, T. K.: J. Mol. Biol. *17,* 503 (1966).

C 46 Cohen, E. H.: J. Food Sci. *31,* 746 (1966).

C 47 Cohen, S., and Porter, R. R.: Biochem. J. *90,* 278 (1964).

C 48 Cole, E. G., and Mecham, D. K.: Anal. Biochem. *14,* 215 (1966).

C 49 Coleman, E. A., Bula, R. J., and Davis, R. L.: Plant Physiol. *41,* 1681 (1966).

C 50 Colover, J.: Chromatogr. and Electroph. Techniques, Vol. II. I. Smith, editor, W. Heinemann Medical Books, London, 1968, p. 413.

C 51 Cotman, C. W., and Mahler, H. R.: Arch. Biochem. Biophys. *120,* 384 (1967).

C 52 Coutinho, H. B., Katchburian, E., and Pearse, A. G. E.: J. Clin. Path. *19,* 617 (1966).

C 53 Craven, G. R., Steers, E., and Anfinsen, C. B.: J. Biol. Chem. *240,* 2468 (1965).

C 54 Croxatto, H. D., Nishimura, E. T., Yamada, K., and Hokama, Y.: J. Immunol. *100,* 563 (1968).

C 55 Cunningham, V. R.: J. Clin. Path. *17,* 143 (1964).

C 56 Cunningham, V. R., and Field, E. J.: J. Neurochem. *11,* 281 (1964).

D 1 Dabbous, M. K., and Draus, F. J.: J. Chromatogr. *35,* 134 (1968).

D 1a Dahlberg, A. E., Dingman, C. W., and Peacock, A. C.: J. Mol. Biol. *41,* 139 (1969).

D 2 Dale, G., and Latner, A. L.: The Lancet *1,* 847 (1968).

D 2a Dale, G., and Latner, A. L.: Clin. Chim. Acta *24,* 61 (1969).

D 2b Daniels, M. J., and Wild, D. G.: Anal. Biochem. *35,* 544 (1970).

D 3 Darcy, D. A.: Clin. Chim. Acta *21,* 161 (1968).

D 4 Davajan, V., and Kunitake, G. M.: Fert. Ster. *19,* 623 (1968).

D 5 Davis, B. J.: Preprint "Disc Electrophoresis", Distillation Prod. Div. Eastman Kodak Co., Rochester, N. Y., 1962.

D 6 Davis, B. J.: Enzyme Analysis *3b,* Canalco (1963).

D 7 Davis, B. J.: Annals N. Y. Acad. Sci. *121,* 404 (1964).

D 8 Davis, B.: Canalco Newsletter No. 11, p. 14 (1969).

D 9 Davis, G.: Canalco Abstract No. 349 (1966).

D 10 Davis, G. M., and Lindsay, G.: Ann. Rep. Amer. Malacological Union (1964), p. 20.

D 11 De Boer, J., and Sarnaker, R.: Med. Proc. South Africa *2,* 218 (1956), cited in Pearse, A. E. G.: Histochemistry, Little, Brown & Co., Boston, 2nd edition. 1961, p. 827.

D 12 Delank, H. W.: Klin. Wschr. *46,* 779 (1968).

D 12a Demus, H., and Mehl, E.: Biochim. Biophys. Acta *203,* 291 (1970).

D 13 DePinto, J. A., and Campbell, L.: Biochemistry *7,* 114 (1968).
D 14 Desborough, S., and Irwin, M. R.: Physiol. Zool. *39,* 66 (1966).
D 15 Desborough, S., and Peloguin, S. J.: Phytochem. *6,* 989 (1967).
D 15a Determann, H.: Protides Biol. Fluids *14,* 563 (1966).
D 16 Dietz, A. A., and Lubrano, T.: Anal. Biochem. *20,* 246 (1967).
D 17 Dietz, A. A., Rubinstein, H. M., and Lubrano, T.: Clin. Chem. *12,* 25 (1966).
D 18 Dike, G. W. R., and Bew, F. E.: J. Clin. Path. *20,* 97 (1967).
D 19 Dilworth, R. H.: Ortec Publ. Life Sci. *1,* No. 2 (1969).
D 20 Dingman, C. W., and Peacock, A. C.: Biochemistry *7,* 659 (1968).
D 21 Dingman, C. W., Aronow, A., Bunting, S. L., Peacock, A. C., and O'Malley, B. W.: Biochemistry *8,* 489 (1969).
D 22 Dittmar, K., Kochwa, S., Zucher-Franklin, D., and Wasserman, L. R.: Blood *31,* 81 (1968).
D 23 Doane, W. W.: Canalco Abstract No. 126 B (1964).
D 24 Doane, W. W.: Drosophila Inform. Service *40,* 97 (1964).
D 25 Doane, W. W.: Drosophila Inform. Service *41,* 192 (1966).
D 26 Doane, W. W.: J. Exptl. Zool. *164,* 363 (1967).
D 27 Doman, E., and Koide, S. S.: Biochim. Biophys. Acta *128,* 209 (1966).
D 27a Domschke, W., Seyde, W., and Domagk, G. F.: Z. klin. Chem. und klin. Biochem. *8,* 319 (1970).
D 28 Dorner, M. M., Yount, W. J., and Kabat, E. A.: J. Immunol. *102,* 273 (1969).
D 29 Dounce, A. L., and Hilgartner, C. A.: Exp. Cell. Res. *36,* 228 (1964).
D 30 Dowding, B., and Tárnoky, A. L.: Proc. Assoc. Clin. Biochem. *4,* 18 (1966).
D 31 Dowding, B., and Tárnoky, A. L.: Proc. Assoc. Clin. Biochem. *4,* 167 (1967).
D 32 Doyle, D., and Laufer, H.: J. Cell Biol. *40,* 62 (1969).
D 33 Drapeau, G. R., and Yanofsky, C.: J. Biol. Chem. *242,* 5434 (1967).
D 34 Dravid, A. R., Fredén, H., and Larsson, S.: J. Chromatog. *41,* 53 (1969).
D 35 Driedger, A., Johnson, L. D., and Marko, A. M.: Canad. J. Biochem. Physiol. *41,* 2507 (1963).
D 36 Dudacek, W. E., and Narayan, K. A.: Biochim. Biophys. Acta *125,* 604 (1966).
D 37 Duesberg, P. H., and Rueckert, R. R.: Anal. Biochem. *11,* 342 (1965).
D 38 Dufour, D., La Fontaine, A., and Lille, S.: Rev. d'Immunologie *29,* 411 (1965).
D 39 Dunker, A. K., and Rueckert, R. R.: J. Biol. Chem. *244,* 5074 (1969).
D 40 Durbin, R. D.: Nature *210,* 1186 (1966).
D 41 Durzan, D. J.: Canad. J. Botany *44,* 359 (1966).
D 42 Durzan, D. J., and Chalupa, V.: Canad. J. Bot. *46,* 417 (1968).

E 1 Edelman, G. M., and Poulik, M. D.: J. exper. Med. *113,* 861 (1961).
E 2 Edward, J. T.: Chemistry and Industry 276 (1958).
E 3 Eichner, D.: Experientia *22,* 620 (1966).
E 3a Eldridge, A. C., Anderson, R. L., and Wolf, W. J.: Arch. Biochem. Biophys. *115,* 495 (1966).
E 4 Elson, E., and Jovin, T. M.: Anal. Biochem. *27,* 193 (1969).
E 5 Epstein, E., Houvras, Y., and Zak, B.: Clin. Chim. Acta *20,* 335 (1968).
E 6 Epstein, E., Wolf, P. L., Horwitz, J. P., and Zak, B.: Amer. J. Clin. Path. *48,* 530 (1967).
E 7 Evans, J. H., and Quick, D. T.: Arch. Neurol. *14,* 64 (1966).
E 8 Evans, J. H., and Quick, D. T.: Clin. Chem. *12,* 28 (1966).

F 1 Fairbanks, G., Levinthal, C., and Reeder, R. H.: Biochem. Biophys. Res. Comm. *20,* 393 (1965).

F 2 Falaschi, A., and Kornberg, A.: J. Biol. Chem. *241,* 1478 (1966).

F 3 Fambrough, D. M., and Bonner, J.: Biochemistry *5,* 2563 (1966).

F 4 Fambrough, D. M., and Bonner, J.: J. Biol. Chem. *243,* 4434 (1968).

F 5 Fambrough, D. M., and Bonner, J.: Biochim. Biophys. Acta *175,* 113 (1969).

F 6 Fambrough, D. M., Fujimura, F., and Bonner, J.: Biochemistry *7,* 575 (1968).

F 7 Fantes, K. H., and Furminger, I. G. S.: Nature *215,* 750 (1967).

F 8 Farmer, R., Turano, P., and Turner, W. J.: J. Chromatog. *24,* 204 (1966).

F 9 Fawcett, J. S.: FEBS Letters *1,* 81 (1968).

F 10 Fawcett, J. C., and Morris, C. J. O. R.: Separation Science *1,* 9 (1966).

F 11 Felberg, N., and Schultz, J.: Anal. Biochem. *23,* 241 (1968).

F 12 Felgenhauer, K.: Biochim. Biophys. Acta *133,* 165 (1967).

F 13 Felgenhauer, K.: Biochim. Biophys. Acta *160,* 267 (1968).

F 13a Felgenhauer, K.: Clin. Chim. Acta *27,* 305 (1970).

F 13b Felgenhauer, K.: Prot. Biol. Fluids *17,* 505 (1970).

F 13c Felgenhauer, K., Weis, A., and Glenner, G. G.: J. Chromatog. *46,* 116 (1970).

F 14 Felgenhauer, K., Bach, S., and Stammler, A.: Klin. Wschr. *45,* 371 (1967).

F 14a Ferguson, K. A.: Metabolism *13,* 21 (1964).

F 15 Ferris, T. G., Easterling, R. E., and Budd, R. E.: Clin. Chim. Acta *8,* 792 (1963).

F 16 Fessler, J. H., and Bailey, A. J.: Biochim. Biophys. Acta *117,* 368 (1966).

F 17 Feuer, H., and Lynch, U. E.: J. Amer. Chem. Soc. *75,* 5027 (1953).

F 18 Fiala, E. S.: Experientia *23,* 597 (1967).

F 19 Finerty, J. F., and Dimopoullos, G. T.: J. Parasit. *54,* 585 (1968).

F 20 Finkelstein, R. A., Sobocinski, P. Z., Atthasampunna, P., and Charunmethee, P.: J. Immunology *97,* 25 (1966).

F 21 Finlayson, J. S., and Mushinski, J. F.: Biochem. Biophys. Acta *147,* 413 (1967).

F 22 Fish, D. C., Dobbs, J. P., and Carter, R. C.: Experientia *27,* 37 (1971).

F 23 Fisher, S., Fletcher, A. P., Alkjaersig, N., and Sherry, S.: J. Lab. Clin. Med. *70,* 903 (1967).

F 24 Fitschen, W.: Biochem. J. *88,* 13 P (1963).

F 25 Fitschen, W.: Immunology *7,* 307 (1964).

F 26 Flatmark, T., and Sletten, K.: J. Biol. Chem. *243,* 1623 (1968).

F 27 Fleischmaier, R., and Krol, S.: J. Invest. Dermatol. *48,* 359 (1967).

F 28 Florini, J. R., and Brivio, R. P.: Anal. Biochem. *30,* 358 (1969).

F 29 Folk, J. E.: J. Biol. Chem. *238,* 3895 (1963).

F 30 Folk, J. E., and Schirmer, E. W.: J. Biol. Chem. *238,* 3884 (1963).

F 31 Fowlks, W. L.: Invest. Ophthalmology *4,* 611 (1965).

F 32 Fox, D. J., Thurman, D. A., and Boulter, D.: Biochem. J. *87,* 29 P (1963).

F 33 Fox, D. J., Thurman, D. A., and Boulter, D.: Phytochemistry *3,* 417 (1964).

F 34 Fox, E. N., and Wittner, M. K.: Proc. Nat. Acad. Sci. *54,* 1118 (1965).

F 35 Fraenkel, G., Hsiao, C., and Seligman, M.: Science *151,* 91 (1966).

F 36 Francois, C., and Glimcher, M. J.: Biochim. Biophys. Acta *97,* 366 (1965).

F 37 Fredrick, J. F.: Phytochemistry *1,* 153 (1962).

F 38 Fredrick, J. F.: Phytochemistry *2,* 413 (1963).

F 39 Fredrick, J. F.: Physiol. Plant. *16,* 822 (1963).
F 40 Fredrick, J. F.: Annals N. Y. Acad. Sci. *121,* 634 (1964).
F 41 Fredrick, J. F.: Phyton *21,* 85 (1964).
F 42 Friesen, H. G.: Endocrin. *83,* 744 (1968).
F 43 Fukasawa, H.: Nature *212,* 516 (1966).
F 44 Fullerton, P. M., and Barnes, J. M.: Brit. J. Ind. Med. *23,* 210 (1966).
F 45 Furano, A. V.: J. Biol. Chem. *241,* 2237 (1966).

G 1 Gabl, F., and Pastner, D.: Clin. Chim. Acta *13,* 753 (1966).
G 2 Gabriel, O., and Wang, S.-F.: Anal. Biochem. *27,* 545 (1969).
G 3 Ganguly, P., and Moore, R.: Clin. Chim. Acta *17,* 153 (1967).
G 4 Garen, A., and Otsuji, N.: J. Mol. Biol. *8,* 841 (1964).
G 5 Geroch, M. E., Richards, E. G., and Davies, G. A.: Eur. J. Biochem. *6,* 325 (1968).
G 6 Gesteland, R. F., and Staehelin, T.: J. Mol. Biol. *24,* 149 (1967).
G 7 Geyer, E., Marghescu, S., and Muller, J. J.: Klin. Wschr. *45,* 717 (1967).
G 8 Gilbert, L. I., and Huddleston, C. J.: J. Insect. Physiol. *11,* 177 (1965).
G 9 Giorgio, N. A., and Tabachnick, M.: J. Biol. Chem. *243,* 2247 (1968).
G 10 Givol, D., Craven, G. R., Steers, E., and Anfinsen, C. B.: Biochim. Biophys. Acta *113,* 120 (1966).
G 11 Glenner, G. G., and Bladen, H. A.: Science *154,* 271 (1966).
G 11a Götz, H., Scheiffarth, F., and Eberl, M.: Z. klin. Chem. u. klin. Biochem. *8,* 306 (1970).
G 12 Gofman, Y. Y.: Biokhimiya *30,* 1160 (1965).
G 13 Goldberg, A. F., Takakura, K., and Rosenthal, R. L.: Nature *211,* 41 (1966).
G 14 Goldberg, E.: Science *139,* 602 (1963).
G 15 Goldberg, E.: Annals N. Y. Acad. Sci. *121,* 560 (1964).
G 16 Goldberg, E.: Arch. Biochem. Biophys. *109,* 134 (1965).
G 17 Goldberg, E.: Science *148,* 391 (1965).
G 18 Goldberg, E.: Science *151,* 1091 (1966).
G 19 Goldberg, E., and Cather, J. N.: J. Cell. & Comp. Physiol. *61,* 31 (1963).
G 20 Goldberger, R. F.: Anal. Biochem. *25,* 46 (1968).
G 21 Goldstein, G., Warner, N. L., and Holmes, M. C.: J. Nat. Cancer Inst. *37,* 135 (1966).
G 22 Gonzales, I. E.: Ann. Mtg. Amer. Soc. Clin. Path. Abstract No. 27 (1964); Canalco Abstract No. 170 (1965).
G 23 Gordon, A. H., in: Laboratory Techniques in Biochem. and Mol. Biol. Vol. 1, part I, Work, T. S., and Work, E., editors, North-Holland Publ. Co. Amsterdam-London, 1969, p. 1–145.
G 24 Gordon, A. H., and Louis, I. N.: Anal. Biochem. *21,* 190 (1968).
G 24a Gorovsky, M. A., Carlson, K., and Rosenbaum, J. L.: Anal. Biochem. *35,* 359 (1970).
G 25 Gottschalk, A.: Annals N. Y. Acad. Sci. *106,* 168 (1963).
G 26 Gould, H.: Biochemistry *5,* 1103 (1966).
G 27 Gould, H.: Biochim. Biophys. Acta *123,* 441 (1966).
G 28 Gould, H., Bonanou, S., and Kanagalingham, K.: J. Mol. Biol. *22,* 397 (1966).
G 29 Gould, H. J., Pinder, J. C., Matthews, H. R., and Gordon, A. H.: Anal. Biochem. *29,* 1 (1969).
G 30 Grabar, P., and Burtin, P., editors: Immuno-electrophoretic Analysis Elsevier Publ. Co., Amsterdam (1964).
G 31 Graham, G. N., and Gillibrand, I. M.: Biochem. J. *101,* 13P (1966).
G 32 Gray, R. H., and Steffensen, D. M.: Anal. Biochem. *24,* 44 (1968).

G 32a GRÄSSLIN, D., WEICKER, H., and BARWICH, D.: Z. klin. Chem. u. klin. Biochem. *8,* 288 (1970).

G 32b GRÄSSLIN, D.: Personal communication (1970).

G 33 GREENBLATT, C. L., and LINCICOME, D. R.: Exper. Parasitology *19,* 139 (1966).

G 34 GREENHOUSE, A. H., and SPECK, L. B.: Amer. J. Med. Sciences *248,* 115/333 (1964).

G 35 GRESSEL, J., and WOLOWELSKY, J.: Anal. Biochem. *22,* 352 (1968).

G 36 GRESSEL, J., and WOLOWELSKY, J.: Anal. Biochem. *24,* 157 (1968).

G 37 GRINDROD, J., and NICKERSON, T. A.: J. Dairy Sci. *50,* 142 (1967).

G 38 GROSS, D.: Analyst *90,* 380 (1965).

G 39 GROSS, J., and NAGAI, Y.: Proc. Nat. Acad. Sci. *54,* 1197 (1965).

G 40 GROSSBACH, U.: Biochim. Biophys. Acta *107,* 180 (1965).

G 41 GROSSBACH, U.: Unpublished results (1966).

G 42 GROSSBACH, U.: Chromosoma (Berl.) *28,* 136 (1969).

G 42a GROSSBACH, U.: Submitted for publication (1970).

G 43 GROSSBACH, U., and WEINSTEIN, I. B.: Anal. Biochem. *22,* 311 (1968).

G 44 GROVES, M. L.: Biochim. Biophys. Acta *100,* 154 (1965).

G 45 GROVES, M. L., and KIDDY, C. A.: Arch. Biochem. Biophys. *126,* 188 (1968).

G 46 GROVES, W. E., DAVIS, F. C., and SELLS, B. H.: Anal. Biochem. *24,* 462 (1968).

G 46a GUILBAULT, G. G., KUAN, S. S., TULLY, J., and HACKNEY, D.: Anal. Biochem. *36,* 72 (1970).

G 47 GUPTA, G. N.: Anal. Chem. *38,* 1356 (1966).

G 48 GURLEY, L. R., and SHEPHERD, G. R.: Anal. Biochem. *14,* 364 (1966).

H 1 HAGMAN, M., in: Pollen Physiology and Fertilization, H. F. Linskens, edit., North Holland Publ. Co., Amsterdam (1964), p. 242.

H 2 HAMAGUCHI, H.: Proc. Japan Acad. *42,* 1099 (1966) and *43,* 332 (1967).

H 3 HAMASHIGE, S., ASTOR, M. A., ARQUILLA, E. R., and VANTHIEL, D. H.: J. Clin. Endocrin. Metab. *27,* 1690 (1967).

H 4 HAMILTON, M. G., and RUTH, M. E.: Biochemistry *6,* 2585 (1967).

H 5 HAMMACK, W. J., MOORE, M., and ALBERT, B.: Annals N. Y. Acad. Sci. *121,* 437 (1964).

H 6 HANNAN, R. S., and LEA, C. H.: Biochim. Biophys. Acta *9,* 293 (1952).

H 7 HANNIG, K.: Z. analyt. Chemie *181,* 244 (1961).

H 7a HANNIG, K., and WIRT, H.: Z. analyt. Chem. *243,* 522 (1968).

H 8 HANOCK, R.: J. Mol. Biol. *40,* 457 (1969).

H 9 HANSL, R.: Annals N. Y. Acad. Sci. *121,* 391 (1964).

H 10 HANSTEIN, A., and MÜLLER-EBERHARD, U.: J. Lab. Clin. Med. *71,* 232 (1968).

H 11 HARTMAN, B. K., and UDENFRIEND, S.: Anal. Biochem. *30,* 391 (1969).

H 12 HASHIMOTO, C., MCSHAN, W. H., and MEYER, R. K.: Biochemistry *5,* 3419 (1966).

H 12a HASSALL, H., LUNN, P. J., and RYALL-WILSON, J.: Anal. Biochem. *35,* 326 (1970).

H 13 HATHAWAY, G., and CRIDDLE, R. S.: Proc. Nat. Acad. Sci. *56,* 680 (1966).

H 14 HAWKER, C. D., GLASS, J. D., and RASMUSSEN, H.: Biochemistry *5,* 344 (1966).

H 15 HEDRICK, J. L., and SMITH, A. J.: Arch. Biochem. Biophys. *126,* 155 (1968).

H 16 HEIDEMAN, M. L.: Annals N. Y. Acad. Sci. *121,* 501 (1964).

H 17 HEIDEMAN, M. L.: Biochemistry *3,* 1108 (1964).

H 18 HEMMINGSEN, L., and OTHER, A.: Acta Ophthalmol. *45,* 359 (1967).

H 19 HENNING, U., and YANOFSKY, C.: J. Mol. Biol. *6,* 16 (1963).

H 20 Herrick, H. E., and Lawrence, J. M.: Anal. Biochem. *12,* 400 (1965).
H 20a Hilborn, J. C., and Anastassiadis, P. A.: Anal. Biochem. *31,* 51 (1969).
H 20b Hill, R. J., and Wake, R. G.: Anal. Biochem. *36,* 521 (1970).
H 21 Hilty, J. W., and Smithhenner, A. F.: Phytopathology *56,* 287 (1966).
H 22 Hindley, J.: J. Mol. Biol. *30,* 125 (1967).
H 23 Hjertén, S.: Arch. Biochem. Biophys. Suppl. *1,* 147 (1962).
H 24 Hjertén, S.: J. Chromatog. *11,* 66 (1963).
H 25 Hjertén, S.: Prot. Biol. Fluids *14,* 553 (1967).
H 26 Hjertén, S., Jerstedt, S., and Tiselius, A.: Anal. Biochem. *11,* 211 (1965).
H 27 Hjertén, S., Jerstedt, S., and Tiselius, A.: Anal. Biochem. *11,* 219 (1965).
H 28 Hjertén, S., Jerstedt, S., and Tiselius, A.: Anal. Biochem. *27,* 108 (1969).
H 29 Hoagland, P. D.: Anal. Biochem. *26,* 194 (1968).
H 30 Hoch, J. A., Simpson, F. J., and Demoss, R. D.: Biochemistry *5,* 2229 (1966).
H 31 Hoffman, H., Naughton, M. A., McDougall, J., and Hamilton, E. A.: Nature *214,* 703 (1967).
H 32 Hoffman, H., and McDougall, J.: Exp. Cell Res. *51,* 485 (1968).
H 33 Hokama, Y., and Riley, R. F.: Biochim. Biophys. Acta *74,* 305 (1963).
H 34 Hokama, Y., Croxatto, H. D., Yamada, K., and Nishimura, E. T.: Cancer Res. *27,* 2300 (1967).
H 35 Holland, N. H., Hong, R., Davis, N. C., and Clark, C. D.: J. Pediatrics *61,* 181 (1962).
H 36 Hollmén, T., and Kulonen, E.: Anal. Biochem. *14,* 455 (1966).
H 37 Holmes, R.: Biochim. Biophys. Acta *133,* 174 (1967).
H 38 Holtzman, N. A., Naughton, M. A., Iber, F. L., and Gaumitz, B. M.: J. Clin. Invest. *46,* 993 (1967).
H 39 Hong, R., and Good, R. A.: Science *156,* 1102 (1967).
H 40 Horstmann, H. J.: Hoppe-Seyler's Z. physiol. Chem. *349,* 131 (1969).
H 41 Hotham-Iglewski, B., Philipp, L. A., and Franklin, R. M.: Nature *218,* 700 (1968).
H 42 Huang, R. C., and Bonner, J.: Proc. Nat. Acad. Sci. *54,* 960 (1965).
H 43 Huang, R. C., and Huang, P. C.: J. Mol. Biol. *39,* 365 (1969).
H 44 Humm, D. G., and Sylvia, A. L.: Science *150,* 635 (1965).
H 45 Huneeus-Cox, F.: Science *143,* 1036 (1964).
H 46 Hunter, R. L., and Maynard, E. A.: J. Histochem. Cytochem. *10,* 677, (1962).
H 47 Hunter, R. L., Rocha, J. T., Pfrender, A. R., and De Jong, D. C.: Annals N. Y. Acad. Sci. *121,* 532 (1964).
H 48 Hutchison, C. A., Edgell, M. H., and Sinsheimer, R. L.: J. Mol. Biol. *23,* 553 (1967).
H 49 Hydén, H., Bjurstam, K., and McEwen, B.: Anal. Biochem. *17,* 1 (1966).
H 50 Hydén, H., and Lange, P. W., in: Macromolecules and the Function of the Neuron; Lodin, Z., and Rose, S. P. R., editors, Excerpta Medica Foundation Amsterdam 1968, p. 33.

I 1 Iandolo, J. J.: Anal. Biochem. *36,* 6 (1970).
I 1a Ingram, L., Tombs, M. P., and Hurst, A.: Anal. Biochem. *20,* 24 (1967).
I 2 Innl, P.: Clin. Chim. Acta *19,* 205 (1968).
I 3 Isichei, U. P.: Med. Thesis, Univ. München (1966).

J 1 Jacey, M. J., and Schaefer, K. E.: Amer. J. Physiol. *212,* 859 (1967).
J 2 Jamieson, G. A., and Gaffney, P. J.: Biochim. Biophys. Acta *154,* 96 (1968).
J 3 Jarabak, J., Seeds, A. E., and Talalay, P.: Biochemistry *5,* 1269 (1966).

J 4 Javid, J.: Proc. Nat. Acad. Sci. *57,* 920 (1967).
J 5 Jeppson, J. O.: Biochim. Biophys. Acta *140,* 468 (1967).
J 6 Jirka, M.: FEBS Letters *4,* 28 (1969).
J 7 John, D. W., and Miller, L. L.: Lab. Invest. *14,* 1402 (1965).
J 7a Johns, E. W.: Biochem. J. *104,* 78 (1967).
J 8 Johns, E. W.: J. Chromatog. *42,* 152 (1969).
J 9 Johnson, B. L.: Science *158,* 131 (1967).
J 10 Johnson, B. L., Barnhart, D., and Hall, O.: Amer. J. Botany *54,* 1089 (1967).
J 11 Johnson, B. L.: Nature *216,* 859 (1967).
J 12 Johnson, D. L., Driedger, A., and Marko, A. M.: Canad. J. Biochem. Physiol. *42,* 795 (1964).
J 13 Johnson, O. W., and Stephen, W. P.: Nature *203,* 207 (1964).
J 14 Jolley, W. B., and Allen, H. W.: Nature *208,* 390 (1965).
J 15 Jones, A. E., Fisher, J. N., Lewis, U. J., and Vanderlaan, W. P.: Endocrinology *76,* 578 (1965).
J 16 Jongkind, J. F., Wisse, J. H., and Bloemendal, H.: Prot. Biol. Fluids *10,* 77 (1963).
J 17 Jordan, E. M., and Raymond, S.: Anal. Biochem. *27,* 205 (1969).
J 18 Jovin, T., Chrambach, A., and Naughton, M. A.: Anal. Biochem. *9,* 351 (1964).
J 19 Juul, P.: Clin. Chim. Acta *19,* 205 (1968).

K 1 Kaback, H. R., and Stadtman, E. R.: Proc. Nat. Acad. Sci. *55,* 920 (1966).
K 2 Kalberer, F., and Rutschmann, J.: Helv. Chim. Acta *44,* 1956 (1961).
K 3 Kalf, G. F., and Gréce, M. A.: Arch. Biochem. Biophys. *107,* 141 (1964).
K 3a Kaltschmidt, E., and Wittmann, H. G.: Anal. Biochem. *36,* 401 (1970).
K 4 Kang, A. H., Nagai, Y., Piez, K. A., and Gross, J.: Biochemistry *5,* 509 (1966).
K 5 Kao, K. J. T., Leslie, J. G., and McGavack, T. H.: Proc. Soc. Exp. Biol. Med. *122,* 1129 (1966).
K 6 Karnovsky, M. J., and Roots, L.: J. Histochem. Cytochem. *12,* 219 (1964).
K 7 Kaschnitz, R.: Z. Klin. Chem. u. Klin. Biochem. *5,* 126 (1967).
K 8 Kates, J. R., and Goldstein, L.: J. Protozool. *11,* 30 (1964).
K 9 Katzman, P. A, and Doisy, E. A.: J. Biol. Chem. *98,* 739 (1932).
K 10 Kellenberger, E.: Virology *34,* 549 (1968).
K 11 Keller, P. J., and Allan, B. J.: J. Biol. Chem. *242,* 281 (1967).
K 12 Keutel, H. J.: Annals N. Y. Acad. Sci. *121,* 484 (1964)
K 13 Keyser, J. W.: Anal. Biochem. *9,* 249 (1965).
K 14 Keyser, J. W.: Anal. Biochem. *12,* 395 (1965).
K 15 Kibler, R. F., Fox, R. H., and Shapira, R:. Nature *204,*, 1273 (1964).
K 15a Kidby, D. K.: Anal. Biochem. *34,* 478 (1970).
K 16 Kierszenbaum, F., Levison, S. A., and Dandliker, W. B.: Anal. Biochem. *28,* 563 (1969).
K 17 Kinard, F. E.: Rev. Sci. Instr. *28,* 293 (1957).
K 17a King, E. E.: J. Chromatog. 53, 559 (1970).
K 18 King, E. R., and Mitchell, T. G.: A Manual for Nuclear Medicine, Thomas Springfield, Ill. (1961).
K 19 King, J. S., Fielden, M. L., and Boyce, W. H.: Proc. Soc. Exper. Biol. Med. *108,* 726 (1962).
K 19a Kingsbury, N., and Masters, C. J.: Anal. Biochem. *36,* 144 (1970).
K 20 Kinkade, J. M., and Cole, R. D.: J. Biol. Chem. *241,* 5790 (1966).
K 20a Kirchner, C.: Wilh. Roux' Archiv *164,* 97 (1969).

K 21 KISCHER, C. W., GURLEY, L. R., and SHEPHERD, G. R.: Nature *212,* 304 (1966).
K 22 KITCHIN, F. D.: Proc. Soc. Exp. Biol. Med. *119,* 1153 (1965).
K 23 KNUDSEN, F. U., and GORMSEN, J.: Scand. J. Hematol. *5,* 361 (1968).
K 23a KNUDSEN, F. U., KNUDSEN, H. E., and GORMSEN, J.: Anal. Biochem. *36,* 192 (1970).
K 24 KOCH, P.: Zeiss-Mitteilungen *4,* 397 (1968).
K 25 KOCHWA, S., SMITH, E., DAVIS, B. J., and WASSERMAN, L. R.: Annals N. Y. Acad. Sci. *121,* 445 (1964).
K 26 KOENIG, E., and BRATTGARD, S. O.: Anal. Biochem. *6,* 424 (1963).
K 26a KOENIG, R., STEGEMANN, H., FRANCKSEN, H., and PAUL, H. L.: Biochim. Biophys. Acta *207,* 184 (1970).
K 27 KOHLRAUSCH, F.: Ann. Physik und Chemie *62,* 209 (1897).
K 28 KOHN, L.: Canalco Newsletter No. 10, p. 9 (1968).
K 29 KONIGSBERG, W., WEBER, K., NOTANI, G., and ZINDER, N.: J. Biol. Chem. *241,* 2579 (1966).
K 30 KONINGS, R. N. H., and BLOEMENDAL, H.: Eur. J. Biochem. *7,* 165 (1969).
K 31 KONYUKHOV, B. V., and WACHTEL, A. W.: Exper. Eye Res. *2,* 325 (1963).
K 31a KOPPERSCHLÄGER, G., DIEZEL, W., BIERWAGEN, D., and HOFMANN, E.: FEBS Letters *5,* 221 (1969).
K 31b Koppikar, S. V., FATTERPAKER, P., and SREENIVASAN, A.: Anal. Biochem. *33,* 366 (1970).
K 32 KORENMAN, S. G., CRAVEN, G. R., and ANFINSEN, C. B.: Biochim. Biophys. Acta *124,* 160 (1966).
K 33 KORNEL, L., SCHROHEN, R. E., and CALDWELL, R. C.: Arch. Biochem. Biophys. *122,* 280 (1967).
K 34 KRAKOW, J. S., DALEY, K., and FRONK, E.: Biochem. Biophys. Res. Commun. *32,* 98 (1968).
K 35 KRAUSE, U., and RAUNIO, V.: Clin. Chim. Acta *17,* 251 (1967).
K 36 KROTOSKI, W. A., and WEIMER, H. E.: Canad. J. Biochem. *45,* 1577 (1967).
K 37 KRUSKI, A. W., and NARAYAN, K. A.: Biochim. Biophys. Acta *168,* 570 (1968).
K 38 KUNITAKE, G. M., NAKAMURA, R. M., WELLS, B. G., and MOYER, D. L.: Fert. Ster. *16,* 120 (1965).
K 39 KURATA, Y.: Stain Techn. *28,* 231 (1953).
K 40 KURNICK, N. B.: Exper. Cell Res. *1,* 151 (1950).
K 41 KURNICK, N. B.: Stain Techn. *30,* 213 (1955).
K 42 KVISTBERG, D., LESTER, G., and LAZAROW, A.: J. Histochem. Cytochem. *14,* 609 (1966).

L 1 LADIŞLAV, C., and DUSA, Z.: Chem. Listy *61,* 1221 (1967).
L 2 LAMY, J., VARGUES, R., and WEILL, J.: Compt. Rend. Soc. Biol. *171,* 1437 (1967).
L 3 LANCHANTIN, G. F., FRIEDMAN, J. A., and HART, D. W.: J. Biol. Chem. *243,* 476 (1968).
L 4 LANCHANTIN, G.: J. Biol. Chem. *244,* 865 (1969).
L 5 LANGDON, R. C., and SLOAN, H. R.: Proc. Nat. Acad. Sci. *57,* 401 (1967).
L 6 LANGVAD, E.: Intern. J. Cancer *3,* 17 (1968).
L 6a LATHE, G. H., and RUTHREN, C. R. J.: Biochem. J. *62,* 665 (1956).
L 7 LAURENT, T. C., and PERSSON, H.: Biochim. Biophys. Acta *78,* 360 (1963).
L 7a LAURENT, T. C., and KILLANDER, J.: J. Chromatog. *14,* 37 (1964).
L 8 LAYCOCK, M. V., THURMAN, D. A., and BOULTER, D.: Clin. Chim. Acta *11,* 98 (1965).

L 9 Leaback, D. H., and Rutter, A. C.: Biochem. Biophys. Res. Commun. *32,* 447 (1968).
L 10 Leach, B. E., Rainey, N. H., and Ballansch, M. J.: J. Dairy Sci. *49,* 1465 (1966).
L 11 Le Bouton, A. V.: Anal. Biochem. *26,* 445 (1968).
L 12 Leboy, P. S., Coy, E. C., and Flaks, J. G.: Proc. Nat. Acad. Sci. *52,* 1367 (1964).
L 13 Lee, J. W.: Biochim. Biophys. Acta *69,* 159 (1963).
L 14 Lee, Y. C., and Sherbaum, O. H.: Biochemistry *5,* 2067 (1966).
L 15 Lees, C. W., and Hartley, R. W.: Biochemistry *5,* 3951 (1966).
L 16 Leise, E. M., Gray, J., and Ward, M. K.: J. Bacteriol.*96,* 154 (1968).
L 17 Lerch, B.: Experientia *24,* 889 (1968).
L 18 Lerch, B., and Stegemann, H.: Anal. Biochem. *29,* 76 (1969)
L 19 Lesuk, A., Terminiello, L., and Traver, J. H.: Science *147,* 880 (1965).
L 20 Leuzinger W., and Baker, A. L.: Proc. Nat. Acad. Sci. *57,* 446 (1967).
L 21 Levy, R. I., and Fredrickson, D. S.: Amer. J. Card. *22,* 576 (1968).
L 21a Lewicki, P. P., and Sinskey, A. J.: Anal. Biochem. *33,* 273 (1970).
L 22 Lewis, U. J.: J. Biol. Cem. *237,* 3141 (1962).
L 23 Lewis, U. J.: J. Biol. Cem. *238,* 3330 (1963).
L 24 Lewis, U. J., and Clark, M. O.: Anal. Biochem. *6,* 303 (1963)
L 25 Lewis, U. J., and Cheever, E. V.: J. Biol. Chem. *240,* 247 (1965).
L 26 Lewis, U. J., Cheever, E. V., and Seavey, B. K.: Anal. Biochem. *24,* 162 (1968).
L 26a Lewis, U. J., Cheever, E. V., and Seavy, B. K.: J. Biol. Chem. *243,* 260 (1968).
L 27 Lewis, U. J., Cheever, E V., and Vanderlaan, W. P.: Endocrinology *76,* 210 (1965).
L 28 Lewis, U. J., Cheever, E. V., and Vanderlaan, W. P.: Endocrinology *76,* 362 (1965).
L 29 Lewis, U. J., Fisher, J. N., and Vanderlaan, W. R.: Acta Chem. Scand. *17,* 165 (1963).
L 29a Lewis, U. J., Litteria, M., and Cheever, E. V.: Endocrinology *85,* 690 (1969).
L 30 Lewisl, U. J., and Vanderlaan, W. R.: J. Biol. Chem. *238,* 3336 (1963).
L 30a Liefänder, M., and Stegemann, H.: Hoppe-Seyler's Z. physiol. Chem. *349,* 157 (1968).
L 31 Lim, R., Huang, J. J., and Davis, G. A.: Anal. Biochem. *29,* 48 (1969).
L 31a Lim, R., and Tadayyon, E.: Anal. Biochem. *34,* 9 (1970).
L 32 Linder, H. J.: Amer. Zool. *4,* 331 (1964).
L 33 Lissitzsky, S., Codaccioni, J.-L., Bismuth, J., and Depieds, R.: J. Clin. Endocrin. *27,* 185 (1967).
L 34 Lockett, M. F., and Siddiqui, H. H.: Brit. J. Pharm. Chemoth. *32,* 311 (1968).
L 35 Loening, U. E.: Biochem. J. *102,* 251 (1967).
L 35a Loening, U. E.: J. Mol. Biol. *38,* 355 (1968).
L 36 Loening, U. E.: Biochem. J. *113,* 131 (1969).
L 37 Loening, U. E., and Ingle, J.: Nature *215,* 363 (1967).
L 37a Loeschke, V., and Stegemann, H.: Z. Naturforschg. *21b,* 879 (1966).
L 38 Logemann, H., in: Methoden der Org. Chemie (Houben-Weyl), Thieme, Stuttgart (1961), Vol. XIV, Part 1, p. 291f.
L 39 Longworth, L. G., in: Electrophoresis, M. Bier, edit., Academic Press, New York (1959), p. 91.
L 40 Louis-Ferdinand, R., and Blatt, W. F.: Clin. Chim. Acta *16,* 259 (1967).
L 41 Low, R. B., and Wool, I. G.: Science *155,* 330 (1967).

L 41a Lowry, O. H., Rosebrough, N. J., Farr, A. L., and Randall, R. J.: J. Biol. Chem. *193,* 265 (1951).
L 42 Lytle, I. M., and Haskell, J. G.: Amer. J. Obstetr. Gynecol. *89,* 156 (1964).
L 43 Lyttleton, J. W.: Biochim. Biophys. Acta *154,* 145 (1968).

M 1 Macko, V., and Stegemann, H.: Hoppe-Seyler's Z. physiol. Chem. *350,* 917 (1969).
M 1a Maier, G. E., and Fischer, R. L.: J. Food Sci. *31,* 482 (1966).
M 1b Maio, J. J., and Schildkraut, C. L.: J. Mol. Biol. *24,* 29 (1967).
M 2 Maizel, J. V.: Annals N. Y. Acad. Sci. *121,* 382 (1964).
M 3 Maizel, J. V.: Science *151,* 988 (1966).
M 3a Makonkawkeyoon, S., and Haque, R.-U.: Anal. Biochem. *36,* 422 (1970).
M 4 Man, E. V., and Whitehead, J. R.: Clin. Chem. *14,* 1002 (1968).
M 5 Mancuso, V. M.: J. Assoc. Offic. Agric. Chem. *47,* 841 (1964).
M 6 Mandel, J. D.: J. Dental Res. *45,* 634 (1966).
M 7 Mandel, J. D., and Katz, R.: J. oral Ther. *4,* 260 (1968).
M 7a Mandel, J. D., Kutscher, A., Denning, C. R., Thompson, R. H., and Zegarelli, E. V.: Amer. J. Dis. Child. *113,* 431 (1967).
M 8 Manganiello, V. C., and Phillips, A. H.: J. Biol. Chem. *240,* 3951 (1965).
M 8a Mann, M. B., and Huang, P. C.: Anal. Biochem. *32,* 138 (1969).
M 9 Marchesis, V. T., and Steers, E.: Science *159,* 203 (1968).
M 10 Marghescu, S.: Arch. Klin. Exp. Dermat. *228,* 327 (1967).
M 11 Margolis, J.: Anal. Biochem. *27,* 319 (1969).
M 11a Margolis, J., and Kenrick, K. G.: Nature *214,* 1334 (1967).
M 11b Margolis, J., and Kenrick, K. G.: Biochem. Biophys. Res. Commun. *27,* 68 (1967).
M 12 Margolis, J., and Kenrick, K. G.: Anal. Biochem. *25,* 347 (1968).
M 13 Margolis, J., and Kenrick, K. G.: Nature *221,* 1056 (1969).
M 14 Margolis, J., and Kenrick, K. G.: Austr. J. Exp. Biol. Med. Sci. *47,* 637 (1969).
M 15 Margolies, M. N., and Goldberger, R. F.: J. Biol. Chem. *241,* 3262 (1966).
M 16 Marinis, S., and Ott, H.: Prot. Biol. Fluids *12,* 420 (1965).
M 17 Markert, C. L., and Hunter, R. L.: J. Histochem. Cytochem. *7,* 42 (1959).
M 18 Marsh, C. L., Jolliff, S. R., and Payne, L. C.: Amer. J. Clin. Path. *41,* 217 (1964).
M 19 Marsh, W. H., Ord, M. G., and Stocken, L. A.: Biochem. J. *93,* 539 (1964).
M 20 Marshall, W. E., and Porath, J.: J. Biol. Chem. *240,* 209 (1965).
M 21 Matioli, G. T., and Niewisch, H. B.: Science *150,* 1824 (1965).
M 22 Matoltsy, A. G.: J. Invest. Dermatol. *42,* 111 (1964).
M 23 Matoltsy, A. G., and Matoltsy, M. N.: J. Invest. Dermatol. *41,* 255 (1963).
M 24 Matson, C. F.: Clin. Chem. *10,* 644 (1964).
M 25 Matson, C. F.: Anal. Biochem. *13,* 294 (1965).
M 26 Mattenheimer, H.: Mikromethoden für das klinisch-chemische und biochemische Laboratorium, Walter de Gruyter, Berlin (1966).
M 27 Maurer, H. R.: Canalco Abstract No. 236 (1965).
M 28 Maurer, H. R.: Z. klin. Chem. *4,* 85 (1966).
M 29 Maurer, H. R.: Hoppe-Seyler's Z. physiol. Chem. *349,* 115 (1968).
M 30 Maurer, H. R.: DGM No. 1, 986, 227 (1968).
M 31 Maurer, H. R.: Unpublished (1969).
M 31a Maurer, H. R., and Dati, F.: Results to be published (1970).
M 32 Mauritzen, C. M., Starbuck, W. C., Saroja, I. S., Taylor, C. W., and Busch, H.: J. Biol. Chem. *242,* 2240 (1967).

M 33 Maynard, E. A.: J. Histochem. Cytochem. *12,* 25 (1964).
M 34 McAllister, jr., H. C., Wan, Y. C., and Irvin, L. J.: Anal. Biochem. *5,* 321 (1963).
M 35 McCallister, B. D., Bayrd, E. D., Harrison, E. G., and McGuckin, W. F.: Amer. J. Med. *43,* 394 (1967).
M 36 McEwen, B. S.: Anal. Biochem. *25,* 172 (1968).
M 37 McEwen, B. S., and Hydén, H.: J. Neurochem. *13,* 823 (1966).
M 38 McGuckin, W. F., and McKenzie, B. F.: Clin. Chem. *4,* 476 (1958).
M 39 McIlwain, H., and Buddle, H. L.: Biochem. J. *53,* 412 (1953).
M 40 McPhie, P., Hounsell, J., and Gratzer, W. B.: Biochemistry *5,* 988 (1966).
M 41 Mehl, E., and Jatzkewitz, H.: Biochem. Biophys. Res. Commun. *19,* 407 (1965).
M 42 Meisinger, M. A. P., Cirillo, V. J., Davis, G. E., and Reisfeld, R. A.: Nature *201,* 820 (1964).
M 43 Mengoli, H. F., and Watne, A. L.: Nature *212,* 481 (1966).
M 44 Metzenberg, R. L.: Arch. Biochem. Biophys. *100,* 503 (1963).
M 45 Metzger, H.: Proc. Nat. Acad. Sci. *57,* 1490 (1967).
M 46 Meyer, T. S., and Lamberts, B. L.: Biochim. Biophys. Acta *107,* 144 (1965).
M 47 Meyer, T. S., and Lamberts, B. L.: Nature *205,* 1215 (1965).
M 48 Michl, H., and Pastuszyn, A.: Mikrochim. Ichnoanal. Acta *5,* 880 (1963).
M 49 Mikolajcik, E. M.: J. Dairy Sci. *51,* 457 (1968).
M 50 Mills, D. R., Peterson, R. L., and Spiegelman, S.: Proc. Nat. Acad. Sci. *58,* 217 (1967).
M 51 Milkman, R., and Gershwin, M. E.: Canalco Abstract No. 238 (1965).
M 52 Mitchell, H. K., and Weber, U. M.: Science *148,* 964 (1965).
M 53 Mitchell, W. M.: Biochim. Biophys. Acta *147,* 171 (1967).
M 54 Mittelbach, F.: Z. ges. exp. Med. *150,* 59 (1969).
M 55 Mittelbach, F., Antl, W., and Hamburger, K.: Z. klin. Chem. u. klin. Biochem. *7,* 189 (1969).
M 56 Mogi, G.: Seibutsu-Butsuri-Kagaku (Japan) *11,* 99 (1965).
M 57 Mogi, G.: Seibutsu-Butsuri-Kagaku (Japan) *11,* 171 (1965).
M 58 Mogi, G., Aoyagi, R., and Nakamura, S.: Proc. Symp. Chem. Physiol. Pathol. (Japan) *5,* 109 (1965).
M 59 Mogi, G., Nakamura, S., and Asagami, Y.: Seibutsu-Butsuri-Kagaku (Japan) *11,* 15 (1965).
M 60 Möller, W., and Chrambach, A.: J. Mol. Biol. *23,* 377 (1967).
M 61 Monseu, G., and Cumings, J. N.: J. Neurol. Neurosurg. Psych. *28,* 56 (1965).
M 62 Montie, T. C., Montie, D. B., and Ajil, S. J.: Arch. Biochem. Biophys. *114,* 123 (1966).
M 63 Moore, R. O., and Villee, C. A.: Science *142,* 389 (1963).
M 64 Morell, A. G., and Scheinberg, I. H.: Science *131,* 930 (1960).
M 65 Morgan, T. E., and Hanahan, D. J.: Biochemistry *5,* 1060 (1966).
M 66 Morris, C. J. O. R.: Protides Biol. Fluids *14,* 543 (1966).
M 67 Morris, C. J. O. R., and Morris, P.: Separation Methods in Biochemistry, Sir Isaak Pitman & Sons Ltd., London (1963), p. 639f.
M 68 Morris, R. S., Spies, J. R., and Coulson, E. J.: Arch. Biochem. Biophys. *110,* 300 (1965).
M 69 Moskowitz, M. S., and Moskowitz, A. A.: Science *149,* 72 (1965).
M 70 Moss, B., and Ingram, V. M.: Proc. Nat. Acad. Sci. *54,* 967 (1965).
M 71 Moss, B., and Ingram, V.: J. Mol. Biol. *32,* 481 (1968).
M 72 Muramatsu, M., and Bush, H.: Anal. Biochem. *4,* 384 (1962).
M 73 Muus, J., and Vnenchak, J. M.: Nature *204,* 283 (1964).

N 1 Nachlas, M. M., Morris, B., Rosenblatt, D., and Seligman, A. M.: J. Biophys. Biochem. Cytol. *7,* 261 (1960).
N 2 Nachman, R. L., and Ferris, B.: J. of Clin. Invest. *47,* 2530 (1968).
N 3 Nagai, Y., Gross, J., and Piez, K. A.: Annals N. Y. Acad. Sci. *121,* 494 (1964).
N 4 Nagy, L. K., Rogerson, B., and Tomkuss, N.: Nature *212,* 923 (1966).
N 5 Nakamura, S.: Saishin-Igaku (Japan) *19,* 827 (1964).
N 6 Nakamura, S., Asagami, Y., and Mogi, G.: Nó to Shinkei (Japan) *17,* 561 (1965).
N 7 Nakamura, S., Mogi, G., Asagami, Y., and Aoyagi, R.: Seikagaku (Japan) *36,* 596 (1964).
N 8 Nakao, A., Davis, W. J., and Roboz-Einstein, E.: Biochim. Biophys. Acta *130,* 163 and 171 (1966).
N 9 Nakao, A., and Roboz-Einstein, E.: Annals N. Y. Acad. Sci. *122,* 171 (1965).
N 10 Narayan, K. A.: Anal. Biochem. *18,* 582 (1967).
N 11 Narayan, K. A., Dudacek, W. E., and Kummerow, F. A.: Biochim. Biophys. Acta *125,* 581 (1966).
N 12 Narayan, K. A., Dudacek, W. E., and Kummerow, F. A.: Clin. Chim. Acta *14,* 797 (1966).
N 13 Narayan, K. A., Greinin, H. L., and Kummerow, F. A.: J. Lipid Res. *7,* 150 (1966).
N 14 Narayan, K. A., and Kummerow, F. A.: Clin. Chim. Acta *13,* 532 (1966).
N 15 Narayan, K. A., Narayan, S., and Kummerow, F. A.: J. Chromatog. *16,* 187 (1964).
N 16 Narayan, K. A., Narayan, S., and Kummerow, F. A.: Nature *205,* 246 (1965).
N 17 Narayan, K. A., Narayan, S., and Kummerow, F. A.: Clin. Chim. Acta *14,* 227 (1966).
N 18 Narayan, K. A., Vogel, M., and Lawrence, J. M.: Anal. Biochem. *12,* 526 (1965).
N 19 Nathans, D., Oeschger, M. P., Eggen, K., and Shimura, Y.: Proc. Nat. Acad. Sci. *56,* 1844 (1966).
N 19a Needles, H. L.: Anal. Biochem. *35,* 533 (1970).
N 20 Neelin, J. M., and Conell, G. E.: Biochim. Biophys. Acta *31,* 539 (1959).
N 21 Neerhout, R. C., Kimmel, J. R., Wilson, J. F., and Lahey, M. E.: J. Lab. Clin. Med. *67,* 314 (1966).
N 22 Neet, K. E., and Friess, S. L.: Arch. Biochem. Biophys. *99,* 484 (1962).
N 23 Neidle, A., and Waelsch, H.: Science *145,* 1059 (1964).
N 24 Neitlich, H. W.: J. Clin. Invest. *45,* 380 (1966).
N 25 Nelson jr., T. E., and Hale, A.: J. Lab. Clin. Med. *68,* 838 (1966).
N 26 Neu, H. C., and Heppel, L. A.: J. Biol. Chem. *240,* 3685 (1965).
N 27 Neuhoff, V.: Arzneimittel-Forschg. (Drug. Res.) *18,* 35 (1968).
N 28 Neuhoff, V., and Lezius, A.: Z. Naturforschg. *23b,* 812 (1968).
N 29 Neuhoff, V., Mühlberg, B., and Meier, J.: Arzneimittel-Forschg. *17,* 649 (1967).
N 30 Neuhoff, V., and Schill, W. B.: Hoppe-Seyler's Z. physiol. Chem. *349,* 795 (1968).
N 31 Neuhoff, V., Schill, W. B., and Sternbach, H.: Hoppe-Seyler's Z. phys. Chem. *349,* 1126 (1968).
N 32 Neuhoff, V., Schill, W. B., and Sternbach, H.: Arzneimittel-Forschg. *19,* 336 (1969).
N 33 Neuhoff, V., Schill, W. B., and Sternbach, H.: Hoppe-Seyler's Z. physiol. Chem. *350,* 335 (1969).

N 34 NEUHOFF, V., SCHILL, W. B., and STERNBACH, H.: Hoppe-Seyler's Z. physiol. Chem. *350,* 767 (1969).
N 35 NEVILLE JR., D. M.: Biochim. Biophys. Acta *133,* 168 (1967).
N 36 NEWCOMBE, D. S., and COHEN, A. S.: Biochim. Biophys. Acta *104,* 480 (1965).
N 37 NILSSON, L.-Ä.: Acta Path. Microb. Scand. *73,* 129 (1968).
N 38 NILSSON, U.: Acta Path. Microb. Scand. *70,* 469 (1967).

O 1 O'CONNOR, R., ROSENBROOK, WM., and ERICKSON, R.: Science *145,* 1320 (1964).
O 2 OGATA, T., and MOGI, G.: Jibiinkoka Rinsho (Japan) *57,* 411 (1964).
O 2a OGSTON, A. G.: Trans. Faraday Soc. *54,* 1754 (1958).
O 3 OH, Y. H., and SANDERS, B. E.: Abal. Biochem. *15,* 232 (1966).
O 4 OLIVERA, B. M., BAINE, P., and DAVIDSON, N.: Biopolymers *2,* 245 (1964).
O 5 OLIVERIO, V. T., DENHAM, C., and DAVIDSON, J. D.: Anal. Biochem. *4,* 188 (1962).
O 6 ORD, M. G., RAAF, J. H., SMIT, J. A., and STOCKEN, L. A.: Biochem J. *95,* 321 (1965).
O 7 ORD, M. G., and STOCKEN, L. A.: Biochem. J. *98,* 888 (1966).
O 8 ORNSTEIN, L.: Enzyme Analysis, C 8, Canalco (1963).
O 9 ORNSTEIN, L.: Annals N. Y. Acad. Sci. *121,* 321 (1964).
O 9a ORNSTEIN, L.: Canalco Newsletter No. 9, p. 8 (1967).
O 10 Ortec 4200 Electrophoresis System (firm 22), Instruction Manual (1969).
O 11 ÖSTNER, V., and HULTIN, T.: Biochim. Biophys. Acta *154,* 376 (1968).
O 12 OTT, H.: Prot. Biol. Fluids *10,* 305 (1963).
O 13 OWEN, J. A., and SMITH, H.: Clin. Chim. Acta *6,* 441 (1961).

P 1 PALVA, T., and RAUNIO, V.: Acta Oto-Laryngol. *63,* 128 (1967).
P 2 PALVA, T., and RAUNIO, V.: Ann. Otol. Rhinol. Laryngol. *76,* 23 (1967)
P 3 PANYIM, S., and CHALKLEY, R.: Arch. Biochem. Biophys. *130.* 337 (1969).
P 4 PANYIM, S., and CHALKLEY, R.: Biochemistry *8,* 3972 (1969).
P 4a PANYIM, S., and CHALKLEY, R.: Biochem. Biophys. Res. Commun. *37,* 1042 (1969).
P 5 PAPADOPOULOS, N. M., and SUTER C. G.: Clin. Chem. *11,* 822 (1965).
P 5a PARISH, C.R., and MARCHALONIS, J. J.: Anal. Biochem. *34,* 436 (1970).
P 5b PASTEWKA, J. V., NESS, A.T., and PEACOCK, A. C.: Anal. Biochem. *35,* 160 (1970).
P 6 PASTEWKA, J. V., Ness, A. T., and PEACOCK, A. C.: Clin. Chim. Acta *14,* 219 (1966)
P 7 PAYNE, W. R.: J. Assoc. Offic. Agric. Chem. *46,* 1003 (1963).
P 8 PEACOCK, A. C., and DINGMAN, C. W.: Biochemistry *6,* 1818 (1967).
P 9 PEACOCK, A. C., and DINGMAN, C. W.: Biochemistry *7,* 668 (1968).
P 10 PEARSE, A. G. E.: Histochemistry, Little Brown & Co., Boston (1960), p. 913.
P 11 PECHÈRE, I. F.: Personal communication (1968).
P 12 PECKHAM, W. D.: J. Biol. Chem. *242,* 190 (1967).
P 13 PERRY, T. O.: Bioscience *13,* 73 (1963).
P 14 PETERSON, J. I.: Anal. Biochem. *25,* 257 (1968).
P 15 PETERSON, R. C.: J. Pharm. Sci. *56,* 1489 (1967).
P 16 PETRAKIS, P. L.: Anal. Biochem. *28,* 416 (1969).
P 17 PETZ, L. D., FINK, D. J., LETZKY, E. A., FUDENBERG, H. H., and MÜLLER-EBERHARD, H. J.: Clin. Invest. *47,* 2469 (1968).
P 18 PHILPOT, F. J., and PHILPOT, J. ST. L.: Biochem. J. *30,* 2191 (1936).
P 19 PIERCE, J. O., and STEMMER, K. L.: Arch. Envir. Health *12,* 190 (1966).
P 20 PIEZ, K. A.: J. Biol. Chem. *239,* 4315 (1964).
P 21 PIHA, R. S., CUÉNOD, M., and WAELSCH, H.: J. Biol. Chem. *241,* 2397 (1966).

P 22 Plummer, T. H., and Hirs, C. H. W.: J. Biol. Chem. *238*, 1396 (1963).
P 23 Polley, M. J., and Müller-Eberhard, H. J.: J. Ex. Med. *128*, 533 (1968).
P 24 Polter, C.: Z. Naturforschg. *22b*, 340 (1967).
P 25 Pons, M. W., and Hirst, G. K.: Virology *34*, 385 (1968).
P 26 Poulik, M. D.: Nature *180*, 1477 (1957).
P 26a Prat, J. P., Lamy, J. N., and Weill, J. D.: Bull. Soc. Chim. Biol. *51*, 1367 (1969),
P 27 Pratt J. J., and Dangerfield, W. G.: Clin. Chim. Acta *23*, 189 (1969).
P 28 Price, W. H., Harrison, H., and Ferebee, S. H.: Annals N. Y. Acad. Sci. *121*, 460 (1964).
P 29 Price, W. H., Harrison, H., and Molenda, J.: Amer. J. Epidem. *83*, 152 (1966).
P 30 Prusík, Z.: J. Chromatogr. *32*, 191 (1968).
P 31 Ptashne, M.: Proc. Nat. Acad. Sci. *57*, 306 (1967).
P 32 Pun, J. Y., and Lombrozo, I.: Anal. Biochem. *9*, 9 (1964).

Q 1 Quantrano, R. S.: Plant Physiol. *43*, 2057 (1968).
Q 2 Quinlivan, W. L. G.: Arch. Biochem. Biophys. *127*, 680 (1968).

R 1 Rabinowitz, Y., and Dietz, A.: Biochim. Biophys. Acta *139*, 254 (1967).
R 2 Rabinowitz, Y., and Dietz, A.: Blood *29*, 182 (1967).
R 3 Rabinowitz, Y., Lubrano, T., Wilhite, B. A., and Dietz, A. A.: Exp. Cell. Res. *48*, 675 (1967).
R 4 Racusen, D.: Nature *213*, 922 (1967).
R 5 Racusen, D., and Calvanico, M.: Anal. Biochem. *7*, 62 (1964).
R 6 Radhakrishnamurthy, B., Chapman, K., and Berenson, G. S.: Biochim. Biophys. Acta *75*, 276 (1963).
R 7 Rapola, J., and Koskimies, O.: Science *157*, 1311 (1967).
R 8 Rauch-Puntigan, H., in: Methoden der Org. Chemie (Houben-Weyl), Thieme, Stuttgart (1961), Vol. XIV, Part 1, p. 1026 f.
R 9 Rausch, W. H., Ludwick, T. M., and Weseli, D. F.: J. Dairy Sci. *48*, 720 (1965).
R 9a Ray, D. K., Troisi, R. M., and Rappaport, H. P.: Anal. Biochem. *32*, 322 (1969).
R 10 Raymond, S.: Clin. Chem. *8*, 455 (1962).
R 11 Raymond, S.: Annals N. Y. Acad. Sci. *121*, 350 (1964).
R 12 Raymond, S., Miles, J. L., and Lee, J. C. J.: Science *151*, 346 (1966).
R 13 Raymond, S., and Nakamichi, M.: Anal. Biochem. *3*, 23 (1962).
R 14 Raymond, S., and Nakamichi, M.: Nature *195*, 697 (1962).
R 15 Raymond, S., and Nakamichi, M.: Anal. Biochem *7*, 225 (1964).
R 16 Raymond, S., and Weintraub, L.: Science *130*, 711 (1959).
R 17 Rechler, M. M.: J. Biol. Chem. *244*, 551 (1969).
R 18 Reeder, R., and Bell, E.: J. Mol. Biol. *23*, 577 (1967).
R 19 Reich, G., and Legutke, H.: Z. Chem. *3*, 436 (1963).
R 20 Reid, B. R., and Cole, R. D.: Proc. Nat. Acad. Sci. *51*, 1044 (1964).
R 21 Reid, M. S., and Bieleski: Anal. Biochem. *22*, 374 (1968).
R 22 Reisfeld, R. A., Dray, S., and Nisonoff, A.: Immunochemistry *2*, 155 (1965).
R 22a Reisfeld, R. A., Inman, J. K., Mage, R. G., and Appela, E.: Biochemistry *7*, 14 (1968).
R 23 Reisfeld, R. A., Lewis, U. J., and Williams, D. E.: Nature *195*, 281 (1962).
R 24 Reisfeld, R. A., Mucilli, A. S., Williams, D. E., and Steelman, S. L.: Nature *201*, 822 (1964).
R 25 Reisfeld, R. A., and Small, P. A.: Science *152*, 1253 (1966).

R 26 Reisfeld, R. A., Williams, D. E., Cirillo, V. J., Tong, G. J., and Brink, N. G.: J. Biol. Chem. *239,* 1777 (1964).

R 27 Rennels, M. L.: Endocrinology *78,* 659 (1966).

R 28 Rennert, O. M.: Nature *213,* 1133 (1967).

R 29 Resmini, P.: Tecnica Molitoria (Italia) *19,* No. 6 (1968).

R 30 Resmini, P., and Volpe, M.: L'Industria Pastaria (Italia) No. 4 (1968).

R 31 Ressler, N., Springgate, R., and Kaufman, J.: J. Chromatog. *6,* 409 (1961).

R 32 Reusser, F.: Arch. Biochem. Biophys. *106,* 410 (1964).

R 33 Ribeiro, L. P., and McDonald, H. K.: J. Chromatog. *10,* 443 (1963).

R 34 Richards, E. G., Coll, J. A., and Gratzer, W. B.: Anal. Biochem. *12,* 452 (1965).

R 35 Richards, E. G., and Gratzer, W. B.: Nature *204,* 878 (1965).

R 36 Richards, E. G., and Gratzer, W. B.: Chromatogr. and Electroph. Techniques, Vol. II, I. Smith, editor, W. Heinemann Medical Books, London (1968), p. 419.

R 37 Richter, G. W.: Lab. Invest. *12,* 1026 (1963).

R 38 Richter, G. W.: Brit. J. Exper. Pathol. *45,* 88 (1964).

R 39 Ridley, S. M., Thornber, J. P., and Bailey, J. L.: Biochim. Biophys. Acta *140,* 62 (1967).

R 40 Riley, R. F., Coleman, M. K., and Hokama, Y.: Clin. Chim. Acta *11,* 530 (1965).

R 41 Riley, R. F., and Coleman, M. K.: J. Lab. Clin. Med. *72,* 714 (1968).

R 42 Ringborg, U., Eghházi, E., Daneholt, B., and Lambert, B.: Nature *220,* 1037 (1968).

R 43 Ritchie, R. F.: J. Maine Med. Assoc. *58,* 15 (1966).

R 44 Ritchie, R. F., Harter, J. G., and Bayles, T. B.: J. Lab. Clin. Med. *68,* 842 (1966).

R 45 Rivers, S. L., and Sundy, M.: Canalco Abstract No. 149 (1964).

R 46 Robbins, E., and Borun, T. W.: Proc. Nat. Acad. Sci. *57,* 409 (1967).

R 47 Robbins, J. H.: Arch. Biochem. Biophys. *114,* 585 (1966).

R 48 Robboy, S. J., and Kahn, R. H.: Endocrinology *78,* 440 (1966).

R 48a Rodbard, D., and Chrambach, A.: Proc. Nat. Acad. Sci. *65,* 970 (1970).

R 49 Rogers, L. J.: Biochim. Biophys. Acta *94,* 324 (1965).

R 50 Rohlfing, S., Brown, E. R., Schwartz, S. O., and Spira, R.: Proc. Soc. Exp. Biol. Med. *131,* 146 (1969).

R 51 Roholt, O. A., and Pressman, D.: Science *153,* 1257 (1966).

R 52 Roholt, O. A., Radzimski, G., and Pressman, D.: J. Exper. Med. *123,* 921 (1966).

R 53 Rosenberg, S. A., and Guidotti, G.: J. Biol. Chem. *243,* 1985 (1968).

R 54 Rosenszajn, L., Epstein, Y., Shoham, D., and Arber, J.: J. Lab. and Clin. Med. *72,* 786 (1968).

R 55 Rosenthaler, J., Guirard, B. M., Chang, G. W., and Snell, E. E.: Proc. Nat. Acad. Sci. *54,* 152 (1965).

R 56 Rothberg, S., and Axilrod, G. D.: Science *156,* 90 (1967).

R 57 Rothman, U., and Lidén, S.: Nature *208,* 389 (1965).

R 58 Rottem, S., and Razin, S.: J. Bacteriol. *94,* 359 (1967).

R 59 Rudolf, K., and Stahman, M. A.: Plant. Physiol. *41,* 389 (1966).

R 60 Rueckert, R. R., and Duesberg, P. H.: J. Mol. Biol. *17,* 490 (1966).

R 61 Russell, F. E., Buess, F. W., and Woo, M. Y.: Toxicon *1,* 99 (1963).

R 62 Russell, F. E., Buess, F. W., Woo, M. Y., and Eventov, R.: Toxicon *1,* 229 (1963).

S 1 Salton, M. R. J., and Schmitt, M. D.: Biochem. Biophys. Res. Commun. *27,* 529 (1967).

S 2 Salton, M. R. J., Schmitt, M. D., and Trefts, P. E.: Biochem. Biophys. Res. Commun. *29,* 728 (1967).
S 3 Sanders, B. E., Small, S. M., Ayers, W. J., Oh, Y. H., and Axelrod, S.: Transact. N. Y. Acad. Sci. *28,* 22 (1965).
S 4 Saphonov, W. J.: cited in Methods in Enzymology Vol. XI, Academic Press (1967), p. 184.
S 5 Sastry, L. V. S., and Virupaksha, T. K.: Anal. Biochem. *19,* 505 (1967).
S 6 Saxena, B. B., and Henneman, P. H.: Biochem. J. *100,* 711 (1966).
S 7 Saxena, B. B., and Rathnam, P.: J. Biol. Chem. *242,* 3769 (1967).
S 8 Scavini, L. M., and Zuloaga, G.: Rev. Invest. Agropec. (Argentina) *3,* 81 (1966).
S 9 Schaefer, J. L.: Amer. J. Med. Technol. *34,* 21 (1968).
S 10 Schenkein, I., Levy, M., and Weiss, P.: Anal. Biochem. *25,* 387 (1968).
S 11 Scher, W., and Jacoby, W. B.: J. Biol. Chem. *244,* 1878 (1969).
S 12 Scheurlen, P. G., Felgenhauer, K., and Pappas, A.: Klin. Wschr. *45,* 419 (1967).
S 13 Schimke, R. T., Berlin, C. M., Sweeney, E. W., and Caroll, W. R.: J. Biol. Chem. *241,* 2228 (1966).
S 14 Schleyer, F., and Schaible, P.: Z. klin. Chem. Biochem. *5,* 32 (1967).
S 15 Schneiderman, L. J.: Biochem. Biophys. Res. Comm. *20,* 763 (1965).
S 16 Schneiderman, L. J., and Junga, J. G.: Biochemistry *7,* 2281 (1968).
S 17 Schoenmakers, J. G. G., Kurstjens, R. M., Haanen, C., and Zilliken, F.: Thrombosis et Diathesis Haemorrhagica *9,* 546 (1963).
S 18 Schoenmakers, J. G. G., Matze, R., Haanen, C., and Zilliken, F.: Biochim. Biophys. Acta *93,* 433 (1964).
S 19 Schonne, E.: Prot. Biol. Fluids *11,* 368 (1964).
S 20 Schram, E.: Arch. Internat. Physiol. Biochim. *72,* 695 (1964).
S 21 Schram, E., and Roosens, H.: Arch. Internat. Physiol. Biochim. *72,* 697 (1964).
S 22 Schrauwen, J. A. M.: J. Chromatog. *15,* 256 (1964).
S 23 Schrauwen, J. A. M.: J. Chromatog. *23,* 180 (1966).
S 24 Schulz, R., Renner, G., Henglein, A., and Kern, W.: Makromolek. Chem. *12,* 20 (1953).
S 25 Schultz, J.: Anal. Biochem. *23,* 241 (1968).
S 26 Schütte, H.-R.: Radioaktive Isotope in der org. Chemie u. Biochemie, Verlag Chemie, Weinheim (1966).
S 27 Schwabe, C.: Anal. Biochem. *17,* 201 (1966).
S 28 Schwabe, C., and Kalnitsky, G.: Biochemistry *5,* 158 (1966).
S 29 Schwartz, A. N., and Zabin, B. A.: Anal. Biochem. *14,* 321 (1966).
S 30 Schwiegk, H., and Turba, F., edit.: Künstl. radioakt. Isotope in Physiologie, Diagnostik und Therapie, Springer, Berlin (1961).
S 31 Sells, B. H., and Davis jr., F. C.: Science *159,* 1240 (1968).
S 32 Seto, J. T., and Hokama, Y.: Annals N. Y. Acad. Sci. *121,* 640 (1964).
S 33 Sgouris, J. T., Storey, R. W., Wolfe, R. W., and Anderson, G. R.: Transfusion *6,* 146 (1966).
S 34 Shapiro, A. L., and Maizel, J. V.: Anal. Biochem. *29,* 505 (1969).
S 35 Shapiro, A. L., Vinuela, E., and Maizel, J. V.: Biochem. Biophys. Res. Commun. *28,* 815 (1967).
S 36 Shapiro, H. D., Miller, K. D., Harris, A. H.: Exp. Mol. Pathology *7,* 362 (1967).
S 37 Shechter, Y., Landan, J. W., and Neucomer, V. D.: J. Invest. Dermat. *52,* 57 (1969).
S 38 Sheff, M. F., Perry, M. B., and Zacks, S. I.: Biochim. Biophys. Acta *100,* 215 (1965).

S 39 SHEPHERD, G. R.: Canalco Abstract No. 148 (1965).
S 40 SHEPHERD, G. R., and GURLEY, L. R.: Anal. Biochem. *14,* 356 (1966).
S 41 SHERIDAN, J. W., KENRICK, K. G., and MARGOLIS, J.: Biochem. Biophys. Res. Comm. *35,* 474 (1969).
S 41a SIEPMANN, R., and STEGEMANN, H.: Naturwissenschaften *54,* 116 (1967).
S 42 SILANO, V., D'ERRICO, A. M., MUNTONI, F., and POCCHIARI, F.: Ann. Ist. Super. Sanità (Italia) *3,* 753 (1967).
S 43 SILANO, V., D'ERRICO, A. M., MICCO, C., MUNTONI, F., and POCCHIARI, F.: J. Assoc. Off. Anal. Chem. *51,* 1213 (1968).
S 44 SIMONS, K., and BEARN, A. G.: Biochim. Biophys. Acta *133,* 499 (1967).
S 44a SKRYRING, G. W., MILLER, R. W., and PURKAYASTHA, V.: Anal. Biochem. *36,* 511 (1970).
S 44b SLAGEL, D. E., WILSON, C. B., and SIMMONS, P. B.: Ann. N. Y. Acad. Sci. *159,* 490 (1969).
S 45 SLATER, G. G.: Federation Proc. *24,* 225 (1965).
S 46 SLATER, G.: Anal. Biochem. *24,* 215 (1968).
S 47 SMALL JR., P. A., REISFELD, R. A., and DRAY, S.: J. Mol. Biol. *11,* 713 (1965).
S 48 SMALL JR., P. A., REISFELD, R. A., and DRAY, S.: J. Mol. Biol. *16,* 328 (1966).
S 48a SMEDS, S.: J. Chromatog. *44,* 148 (1969).
S 49 SMITH, A. P., and VARON, S.: Biochemistry *7,* 3259 (1968).
S 50 SMITH, E. W., and EVATT, B. L.: J. Lab. Clin. Med. *69,* 1018 (1967).
S 51 SMITH, I., PERRY, J. D., and LIGHTSTONE, P. J.: Clin. Chim. Acta *25,* 17 (1969).
S 52 SMITH, I., and WEISS, J. B.: J. Chromatog. *28,* 494 (1967).
S 53 SMITH, I., editor: Chromatographic and Electrophoretic Techniques, Vol. II: Zone Electrophoresis, W. Heinemann Med. Books Ltd., London (1968), p. 365–496.
S 54 SMITH, J. K., and MOSS, D. W.: Anal. Biochem. *25,* 500 (1968).
S 55 SMITH, J. K., and MOSS, D. W.: Biochem. J. *109,* 44P (1968).
S 56 SMITH, W. F. S. R.: Canalco Abstract No. 423 (1966).
S 57 SMITHIES, O.: Biochem. J. *61,* 629 (1955).
S 58 SMITHIES, O.: Arch. Biochem. Biophys. Suppl. *1,* 125 (1962).
S 59 SODEMAN, W. A.: Amer. J. Trop. Med. Hyg. *16,* 591 (1967).
S 60 SODEMAN, W. A., and MEUWISSEN, J. H. E. T.: J. Parasit. *52,* 23 (1966).
S 61 SONENBERG, M., KIKUTANI, M., FREE, C. A., NADLER, A. C., and DELLACHA, J. M.: Annals N. Y. Acad. Sci. *148,* 532 (1968).
S 61a SPALDING, J., KAJIWARA, K., and MUELLER, G. C.: Proc. Nat. Acad. Sci. *56,* 1535 (1966).
S 62 SPEAR, P. G., and ROIZMAN, B.: Anal. Biochem. *26,* 197 (1968).
S 63 STAPLES, R. C., MCCARTHY, W. J., and STAHMANN, M. A.: Science *149,* 1248 (1965).
S 64 STAPLES, R. C., and STAHMANN, M. A.: Science *140,* 1320 (1963).
S 65 STEERS, E., CRAVEN, G. R., and ANFINSEN, C. B.: J. Biol. Chem. *240,* 2478 (1965).
S 66 STEERS, E., CRAVEN, G. R., and ANFINSEN, C. B.: Proc. Nat. Acad. Sci. *54,* 1174 (1965).
S 66a STEGEMANN, H.: Hoppe-Seyler's Z. physiol. Chem. *348,* 951 (1967).
S 66b STEGEMANN, H.: Z. analyt. Chem. *243,* 573 (1968).
S 67 STEINER, J. C., and KELLER, P. J.: Arch. Oral Biol. *13,* 1213 (1968).
S 68 STELLWAGEN, R. H., and COLE, R. R.: J. Biol. Chem. *243,* 4456 (1968).
S 69 STELOS, P., YAGI, Y., and PRESSMANN, D.: J. Immunology *93,* 106 (1964).
S 70 STEVENSON, G. T., and STRAUS, D.: Biochem. J. *108,* 375 (1968).
S 71 STEWARD, F. C., and BARBER, J. T.: Annals N. Y. Acad. Sci. *121,* 525 (1964).
S 72 STEWARD, F. C., and CHANG, L. O.: J. Exper. Botany *14,* 379 (1963).

S 73 STEWARD, F. C., LYNDON, R. F., and BARBER, J. T.: Amer. J. Botany *52,* 155 (1965).
S 74 ST. GROTH, F. S., DE, WEBSTER, R. G., and DATYNER, A.: Biochim. Biophys. Acta *71,* 377 (1963).
S 75 STRAUCH, L.: Prot. Biol. Fluids *15,* 535 (1967).
S 76 STRICKLAND, R. D., PODLESKI, T. R., GURULE, F. T., FREEMAN, M. L., and CHILDS, W. A.: Anal. Chem. *31,* 1408 (1959).
S 77 STRUIJK, C. B., and BEERTHUIS, R. K.: Biochim. Biophys. Acta *116,* 12 (1966).
S 78 STUYVESANT, V. W.: Nature *214,* 405 (1967).
S 79 SUGIYAMA, T., and NAKADA, D.: J. Mol. Biol. *31,* 431 (1968).
S 80 SULITZEANU, D., and GOLDMAN, W. F.: Nature *208,* 1120 (1965).
S 81 SULITZEANU, D., SLAVIN, M., and YECHESKELI, E.: Anal. Biochem. *21,* 57 (1967).
S 82 SUMMERS, D. F., MAIZEL JR., J. V., and DARNELL JR., J. E.: Proc. Nat. Acad. Sci. *54,* 505 (1965).
S 83 SUNDARAM, T. K., and FINCHAM, J. R. S.: J. Mol. Biol. *10,* 423 (1964).
S 84 SUSSMAN, H. H., SMALL, P. A., and COTLOVE, E.: J. Biol. Chem. *243,* 160 (1968).
S 85 SWEENEY, S. C.: J. Dental Res. *46,* 1171 (1967).

T 1 TABER, H. W., and SHERMAN, F.: Annals N. Y. Acad. Sci. *121,* 600 (1964).
T 2 TAKAHASHI, T., RAMACHANDRAMURTHY, P., and LIENER, I. E.: Biochim. Biophys. Acta *133,* 123 (1967).
T 3 TAKAYAMA, K., MACLENNAN, D. H., TZAGOLOFF, A., and STONER, C. D.: Arch. Biochem. Biophys. *114,* 223 (1966).
T 4 TAPPAN, D. V., JACEY, M. J., and BOYDEN, H. M.: Annals N. Y. Acad. Sci. *121,* 589 (1964).
T 5 TARNOKY, A. L., in: Chromatogr. and Electroph. Techn., Vol. II, I. Smith, editor, W. Heinemann Med. Books, London (1968), p. 389.
T 6 TARNOKY, A. L., and DOWDING, B.: Clin. Chem. *1,* 48 (1967).
T 7 TERRY, W. D., SMALL, P. A., and REISFELD, R. A.: Science *152,* 1628 (1966).
T 8 TERZI, M., and LEVINTHAL, C.: J. Mol. Biol. *26,* 525 (1967).
T 9 THELANDER, L.: J. Biol. Chem. *242,* 852 (1967).
T 10 THORUN, W., and MEHL, E.: Biochim. Biophys. Acta *160,* 132 (1968).
T 11 THORUN, W.: Z. klin. Chem. u. klin. Biochem. *9,*3 (1971).
T 12 THORUP, O. A., STROLE, W. B., and LEAVELL, B. S.: J. Lab. Clin. Med. *58,* 122 (1961).
T 13 TICHY, H.: Anal. Biochem. *17,* 320 (1966).
T 14 TISELIUS, A., HJERTÉN, S., and JERSTEDT, S.: Arch. gesamte Virusforschg. *17,* 512 (1965)
T 15 TISHLER P. V. , and EPSTEINC. J.: Anal. Biochem. *22,* 89 (1968).
T 16 TOMBS, M. P.: Anal. Biochem. *13,* 121 (1965)
T 17 TOMBS, M. P., in: Chromatog. Electroph. Techn., Vol. II, I. Smith, editor, W. Heinemann Med. Books, London (1968), p. 443.
T 18 TOMBS, M. P., in: Chromatogr. Electroph. Techn., Vol. II, I. Smith, editor, W. Heinemann Med. Books, London (1968), p. 452.
T 18a TOMBS, M. P.: Biochem. J. *96,* 119 (1965).
T 19 TONO, H., and KORNBERG, A.: I. Biol. Chem. *242,* 2375 (1967).
T 20 TRAUT, R. R.: J. Mol. Biol. *21,* 571 (1966).
T 21 TRAUT, R. R., MOORE, P. B., DELIUS, H., NOLLER, N., and TISSIÈRES, A.: Proc. Nat. Acad. Sci. *57,* 1294 (1967).
T 22 TROMMSDORFF, E., and KRETZ, R., in: Methoden der org. Chemie (Houben-Weyl), Thieme, Stuttgart (1961), Vol. XIV, Part 1, p. 1017 f.

U 1 Uriel, J.: Bull. Soc. Chim. Biol. *48,* 969 (1966).
U 2 Uriel, J., and Berges, J.: Compt. Rend. *262,* 164 (1966).

V 1 Van de Woude, G. F., and Bachrach, H. L.: Arch. ges. Virusforschg. *23,* 362 (1968).
V 2 Vaughan, J. G., Waite, A., Boulter, D., and Waiters, S.: Nature *208,* 704 (1965).
V 3 Veis, A., and Anesey, J.: J. Biol. Chem. *240,* 3899 (1965).
V 4 Vinuela, E., Algranati, I. D., and Ochoa, S.: Europ. J. Biochem. *1,* 3 (1967).
V 5 Vinuela, E., Salas, M., and Ochoa, S.: Proc. Nat. Acad. Sci. *57,* 729 (1967).
V 6 Vito, E. de, and Santomé, J. A.: Experientia *22,* 124 (1966).
V 7 Vollmert, B.: Grundlagen der makromolekularen Chemie, Springer, Berlin (1962), p. 388f.
V 8 Vos, J., and Van Der Helm, H. J.: J. Neurochem. *11,* 209 (1964).

W 1 Wael, J. de, and Wegelin, E.: Rec. trav. chim. *71,* 1035 (1952).
W 2 Wales, E. E., Englert, E., Winward, R. T., Maxwell, J. G., and Stevens, L. E.: Proc. Soc. Exp. Biol. Med. *132,* 146 (1969).
W 3 Wang, C. H., and Willis, D. L.: Radiotracer Methodology in Biological Science, Prentice-Hall Inc., Englewood Cliffs, N. J. (1965).
W 4 Watanabe, Y., Prevec, L., and Graham, A. F.: Proc. Nat. Acad. Sci. *58,* 1040 (1967).
W 4a Watkin, J. E., and Miller, R. A.: Anal. Biochem. *34,* 424 (1970).
W 5 Weber, K., and Osborn, M.: J. Biol. Chem. *244,* 4406 (1969).
W 5a Wein, J.: Anal. Biochem. *31,* 405 (1969).
W 6 Weinbaum, G., and Markman, R.: Biochim. Biophys. Acta *124,* 207 (1966).
W 6a Weinberg, R. A., Loening, U., Willems, M., and Penman, S.: Proc. Nat. Acad. Sci. *58,* 1088 (1967).
W 7 Weinstein, D., and Douglas, J. F.: Anal. Biochem. *6,* 474 (1963).
W 8 Weiss, A. H., and Christoff, N.: Transact. Amer. Neurol. Assoc. *271* (1964).
W 9 Weiss, A. H., and Christoff, N.: Arch. Neurol. *14,* 100 (1966).
W 10 Weiss, A. H., Smith, E., Christoff, N., and Kochwa, S.: J. Lab. Clin. Med. *66,* 280 (1965).
W 11 Weiss, H. J., and Kochwa, S.: J. Lab. Clin. Med. *71,* 153 (1968).
W 12 Weissmann, S. M., Henry, P., Karon, M., and Wynngate, A.: J. Lab. Clin. Med. *66,* 757 (1966).
W 13 Wenzel, M., and Schulze, P. H.: Tritium-Markierung, Walter de Gruyter, Berlin (1962).
W 14 Whipple, H. E., editor: "Gel electrophoresis", Annals N. Y. Acad. Sci. *121,* 305–650 (1964).
W 15 Whitaker, J. R.: Paper Chromatography and Electrophoresis, Vol. I, Academic Press (1967), p. 147.
W 16 White, M. L.: J. Phys. Chem. *64,* 1563 (1960).
W 17 White, M. L., and Dorion, G. H.: J. Polymer Sci. *55,* 731 (1961).
W 18 Wiedemann, E., in: Hoppe-Seyler/Thierf., Hdb. d. physiol. u. pathol.-chem. Analyse, Springer, Berlin (1953), Vol. 1, p. 54.
W 19 Wieme, R. J.: Annals N. Y. Acad. Sci. *121,* 366 (1964).
W 20 Williams, D. E., and Reisfeld, R. A.: Annals N. Y. Acad. Sci. *121,* 373 (1964).
W 20a Wilson, C. W.: Anal. Biochem. *31,* 506 (1969).
W 21 Wimer, D. C.: Anal. Chem. *30,* 77 (1958).
W 22 Wirt, H., and Hannig, K.: Exposition ACHEMA Frankfurt (Germany, 1967).

W 23 Wisse, J. H.: Biochem. J. *99,* 179 (1966).
W 24 Wisse, J. H., Zweers, A., Jongkind, J. F., Bont, W. S., and Bloemendal, H.: Biochem. J. *99,* 179 (1966).
W 25 Woeller, F. H.: Anal. Biochem. *2,* 508 (1961).
W 25a Wolf, B., Michelin-Lausarot, P., Lesnaw, J. A., and Reichmann, M. E.: Biochim. Biophys. Acta *200,* 180 (1970).
W 26 Wolf, G.: Isotopes in Biology, Academic Press (1964).
W 27 Wolf, G.: Experientia *24,* 890 (1968).
W 28 Woods, E. F.: J. Biol. Chem. *242,* 2859 (1967).
W 29 Woodworth, R. C., and Clark, L. G.: Anal. Biochem. *18,* 295 (1967).
W 30 Work, T. S.: J. Mol. Biol. *10,* 544 (1964).
W 31 Woychik, J. H.: Arch. Biochem. Biophys. *109,* 542 (1965).
W 32 Woychik, J. H., Kalan, E. B., and Noelken, M. E.: Biochemistry *5,* 2276 (1966).
W 33 Wrigley, C. W., Webster, H. L., and Turner, J. F.: Nature *209,* 1133 (1966).
W 34 Wrigley, C. W.: J. Chromatog. *36,* 362 (1968).
W 35 Wrigley, C. W.: Science Tools *15,* 17 (1969).
W 36 Wright, G. L., and Mallmann, W. L.: Proc. Soc. Exp. Biol. Med. *123,* 22 (1966).
W 37 Wright, G. L.: Canalco Newsletter No. 10, p. 6 (1968).
W 38 Wuhrmann, F., and Märki, H. H.: Dysproteinämien und Paraproteinämien, Schwabe & Co., Basel-Stuttgart (1963).

Y 1 Yagi, Y., Maier, P., and Pressmann, D.: Science *147,* 617 (1965).
Y 2 Yanai, R., and Nagasawa, H.: Proc. Soc. Exp. Biol. Med. *131,* 167 (1969).
Y 3 Yoshida, A., and Freese, E.: Biochem. Biophys. Acta *99,* 56 (1965).
Y 4 Young, R. W., and Fulhorst, H. W.: J. Cell Biol. *23,* 104A (1964).
Y 5 Young, R. W., and Fulhorst, H. W.: Anal. Biochem. *11,* 389 (1965).
Y 6 Yu, R. J., Harmon, S. R., and Black, F.: J. Bacter. *96,* 1435 (1968).
Y 7 Yunis, A. A., and Arimura, G. K.: Biochim. Biophys. Acta *118,* 335 (1966).

Z 1 Zacharius, R. M., Zell, T. E., Morrison, J. H., and Woddlock, J. J.: Anal. Biochem. *30,* 148 (1969).
Z 1a Zaitlin, M., and Hariharasubramanian, V.: Anal. Biochem. *35,* 296 (1970).
Z 2 Zeidman, I., Dempsey, I. W., and Shelley, P. B.: Arch. Path. *85,* 481 (1968).
Z 3 Zeldin, M. H., and Ward, J. M.: Nature *198,* 389 (1963).
Z 4 Zingale, S. B., Cedrato, A. E., Viscardi, E. B., and Mattioli, C. A.: Argent. Soc. Clin. Inv. Abstract No. 1 (1964); Canalco Abstract No. 199 (1965).
Z 5 Zingale, S. B., and Mattioli, C. A.: Medicina (Buenos Aires) *21,* 121 (1961).
Z 6 Zingale, S. B., Mattioli, C. A., Bonner, H. D., and Bueno, M. P.: Blood *22,* 152 (1963).
Z 7 Zwaan, J.: Anal. Biochem. *21,* 155 (1967).
Z 8 Zwisler, O., and Biel, H.: Prot. Biol. Fluids *12,* 433 (1965).
Z 9 Zwisler, O., and Biel, H.: Beringwerk-Mitteilungen *46,* 129 (1966).
Z 10 Zwisler, O., and Biel, H.: Z. klin. Chem. *4,* 58 (1966).

Note:

The designations "Canalco Abstract" and "Canalco Newsletter" refer to a literature compilation published by the firm Canal Industrial Corporation (firm 5) since 1962.

List of Firms (Manufacturers)

Firm 1
Amersham/Searle Co.
2000 Nuclear Dr.
Des Plaines, Ill. 60018, USA

Firm 2
Beckmann Instruments Inc.,
2500 Harbor Blvd.
Fullerton, Cal., USA

Firm 3
Bio-Rad Laboratories,
32nd & Griffin Ave.,
Richmond, Cal., USA

Firm 4
Buchler Instruments Inc.,
1327, 16th Street
Fort Lee, N. J., USA

Firm 5
Canal Industrial Co.,
5635, Fisher Lane,
Rockville, Md., USA

Firm 6
C. Desaga, GmbH, Nachf. E. Fecht,
Maaßstr. 26, Postfach 407
69 Heidelberg, Germany

Firm 7
Drummond Scientific Co.,
500, Parkway
Broomall, Pa., USA

Firm 7a
Eastman Kodak Co.,
Eastman Organic Chemicals
Rochester N. Y. 14650, USA

Firm 8
E–C Apparatus Co.,
222 South 40th Street,
Philadelphia, Pa., USA

Firm 9
Eppendorf Gerätebau, Netheler
und Hinz GmbH,
Barkhausenweg 1, Postfach 324,
2 Hamburg 63, Germany

Firm 10
Fluka AG, Chemische Fabrik,
Postfach
CH–9470 Buchs, SG, Switzerland

Firm 11
Gelman Instrument Co.,
600 S. Wagner Rd.,
Ann Arbor, Mich. 48106, USA

Firm 12
Gilford Instrument Labs. Inc.,
132 Artina St.,
Oberlin, Ohio 44074, USA

Firm 13
G. T. Gurr Ltd.,
136/144 New Kings Road,
London S. W. 6, Great Britain

Firm 14
Hamilton Co. Inc.
P. O. Box 307,
Whittier, Cal., USA

Firm 15
Ing. H. Hölzel,
Bernöderweg 7,
825 Dorfen, Germany

Firm 16
Hopkin and Williams Ltd.,
P. O. Box,
Chadwell Heath, Essex, Great Britain

Firm 17
Joyce, Loebl & Co. Ltd.,
Princesway,
Team Valley, Gateshead 11, Great Britain

Firm 18
La Pine Scientific Co.,
6001, S. Knox Ave.,
Chicago 29, Ill., USA

Firm 19
LKB-Producter AB,
16125 Bromma 1, Sweden

Firm 20
E. Merck, A. G.,
Postfach 4119
61 Darmstadt 2, Germany

Firm 21
Mickle Laboratory Engineering Co.,
Mill Works,
Gomshall, Surrey, Great Britain

Firm 22
Ortec, Inc.,
1000 Midland Rd.,
Oak Ridge, Tenn. 37830, USA

Firm 23
Oxoid Ltd.,
20 Southwark Bridge Rd.,
London S. E. 1, Great Britain

Firm 24
Packard Instrument Co. Inc.,
P. O. Box 428
La Grange, Ill., USA

Firm 25
Photovolt Co.,
1115 Broadway,
New York 10, N. Y., USA

Firm 26
N. V. Pleuger, S. A.,
Turnhoutsebaan 511/529,
Wijnegem, Belgium

Firm 27
Quickfit and Quartz, Ltd.
P. O. Box,
Stone, Staffordshire, Great Britain

Firm 28
Savant Instruments Inc.,
221 Park Ave.,
Hicksville, N. Y. 11801, USA

Firm 29
E. Schütt jr.,
Postfach 248
34 Göttingen, Germany

Firm 30
Serva-Entwicklungslabor,
Römerstr. 118, Postfach 1505,
69 Heidelberg, Germany

Firm 31
Shandon Scientific Co. Ltd.,
65 Pound Lane,
Willesden, London N. W. 10, Great Britain

Firm 32
Townson & Mercer (Distributors) Pty. Ltd.,
318 Burns Bay Rd.,
Lane Cove, N. S. W., Australia 2066

Firm 33
Trubore Square Tubing
Ace Glass, Inc.
P. O. Box 688
Vineland, N. J., USA

Firm 34
Wissenschaftlich-technische Werkstätten GmbH.,
Trifthofstr.
8120 Weilheim/Obb., Germany

Firm 35
Carl Zeiss,
Postfach
7082 Oberkochen, Germany

List of Abbreviations:

Bis	N,N'-methylenebisacrylamide
C	degree of cross-linkage in percent (see page 1)
Dimethyl-POPOP	1,4 – bis – 2 – (4-methyl-5-phenyloxazolyl)-benzene
EDTA	ethylenediaminetetraacetic acid, disodium salt
DMPN	3-dimethylaminopropionitrile
Persulfate	ammoniumpersulfate, $(NH_4)_2S_2O_8$
PPO	2,5-diphenyloxazol
SDS	sodium dodecylsulfate
T	total monomer concentration of the gel in percent (see page 1)
TEMED	N,N,N',N'-tetramethylethylenediamine
Tris	tris (hydroxymethyl) aminomethane

Subject Index